MCWP 3-35.1

Cold Weather Operations

U.S. Marine Corps

Coordinating Draft June 2000

DEPARTMENT OF THE NAVY
Headquarters United States Marine Corps
Washington, DC 20380-0001

XX June 2000

FOREWORD

1. Purpose. Marine Corps Warfighting Publication (MCWP) 3-35.1, *Cold Weather Operations,* will provide broad doctrine and tactics, techniques, and procedures (TTP) for commanders and staffs throughout the MAGTF.

2. Scope. This publication illustrates how the MAGTF can be organized, trained and equipped to conduct operations in the cold weather environment. It discusses the operational concepts for cold weather operations across the range of military operations. This doctrine spans planning considerations and TTP for operations at the MEF level through MEU (SOC) operations.

3. Supersession. None

4. Certification. Reviewed and approved this date.

BY DIRECTION OF THE COMMANDANT OF THE MARINE CORPS

J. E. RHODES
Lieutenant General, U.S. Marine Corps
Commanding General
Marine Corps Combat Development Command
Quantico, Virginia

DISTRIBUTION: 000 000000 00

CONTENTS
MCWP 3-35.1

CHAPTER 1
HISTORICAL PERSPECTIVES ON COLD WEATHER OPERATIONS

1001. The White Death

The impact of cold weather on military forces is a brutal and tragic story that has been recorded throughout time. In ancient times, the stories are told of Xenophon and Hannibal's armies that were decimated while marching through the high mountains of Greece and the Italian Alps respectively. In more recent times, George Washington's Continental Army lost unknown numbers of men to cold weather injuries at Valley Forge in the winter of 1777-78. Napoleon lost 250,000 men to cold weather injuries and deaths during the Russian Campaign of 1812. The twentieth century, with statistical data, shows a more frightful picture of cold weather operations. During the winter of 1941-42 on the Eastern Front, German military losses during a 60-day period from cold weather operations were 100,000 casualties; 15,000 of the cold weather injuries resulted in amputations. The American Race to the Rhine in 1944-45 tells a similar story. Placing more emphasis on ammunition and fuel to sustain the momentum, General Omar Bradley, Commanding General of the 12th Army Group, consciously placed cold weather clothing, particularly boots, on a lower priority to the aforementioned. The result was numerous cases of frostbite and trench foot. With 90 percent of all cases in the infantry, they had an average hospital stay of 87 days, half requiring treatment in the United States, with 2 percent of all cases ever returning to combat. The United States suffered 84,000 casualties to cold weather injuries. Not six years later, the US military experienced cold weather operations in Korea. The result was cold weather injuries accounting for 10 percent of all casualties, however most of these casualties occurred during a 3-month period. Cold-wet injuries confronted the British during the Falkland Islands campaign. These injuries became an increasing problem and an operational concern which could have been decisive against the British had that war been prolonged. Russian military operations in Grozny were hampered when logistical vehicles failed to supply augmented food rations to combat units. This logistical breakdown resulted from the inability of vehicles to traffic the mountain roads as they thawed and turned to mud. Thus, to this point in time, cold weather remains a formidable obstacle to be overcome when conducting military operations.

1002. Napoleon's Attack on Russia

Napoleon attacked Russia in 1812 with a combined force (the French and European allies) exceeding 600,000 troops. When his army reached Moscow in late September, it was reduced to 250,000 due to the scorched earth policy and desertion. Napoleon took Moscow in September and waited for the Russians to surrender. His lines of communication (LOC) severely extended Napoleon withdrew to Poland two months later. The cold exerted its influence. Morale shattered, desertion rampant, Napoleon arrived in Poland with less than 10,000 effective. Napoleon called General Kutuzov, Commander of the Russian Imperial Forces, *The Sly Old Fox of the North* because Kutuzov considered winter his staunch ally and called it *General Winter*. Although several major battles were fought, Russian forces never decisively engaged the French. Many accused Kutuzov of indecision, even cowardice, for refusing to engage Napoleon in decisive battle. His attitude toward the matter was to let the weather destroy the enemy for him. He stated, " Our young hotheads are angry with the old man for curbing their desire. They do not reflect that circumstances alone are achieving more than our weapons." French losses in the Campaign of 1812 amounted to half a million men, 160,000 horses and 1,000 guns left behind in Russia. Napoleon himself attributed the defeat to the rigors of the climate. The Tsar Alexander upheld his view speaking of Kutuzov's triumph: "The old fellow ought to be contended. The cold weather has rendered him splendid service."

1003. The Finnish-Soviet War

The 1939 *Soviet Manual for Winter Operations* stated, "The Red Army possesses all the advantages over the armies of the other states in relation to practice and ability to operate in the harsh conditions of the winter period. The advantages flow from the geographic conditions of the USSR with its cold climate belt, from the rich military-historical experience and better equipment of the Red Army for winter operations." Despite all these advantages, the Soviets took a real drubbing at the hands of *General Winter* during the war with Finland in 1939-40. The Finns trained and prepared to fight under the most arduous winter conditions. The Soviets expected to run over the Finland in 8 to 10 days. They found themselves bogged down on one of the worst winters since 1828. For 105 days the world gaped at the Red Army's debacle in Finland. How could the Red Army fall to an army the size of Finland's? They enjoyed a superiority of 40 to 1 in personnel, 30 to 1 in aircraft and 100 to 1 in tanks and artillery. The Stalin purges of the 1930's had undoubtedly degraded the Red Army's command structure and many personnel had not received training before combat. The Red Army's supply system completely broke down. Lacking food, winter clothing and shelter, many Soviet soldiers froze to death. Because the Soviet forces operated mainly along the roads, the Finns were free to operate in the dense forests along the roads. Possessing excellent skiing ability, the Finns would conduct attacks lasting only a few minutes, never attacking armor or heavy artillery but supply trains and field kitchens. These attacks would often occur at night when the Soviets gathered around warming fires. By concentrating attacks on the combat service support elements of the invasion divisions, the Finns brought the divisions to a halt. Once this occurred, the Finns proceeded to breakup the enemy division, denying the Soviets the ability to transport rations, fuel and ammunition to the isolated groups. Better known as *motti* tactics, this resulted in the complete destruction of the 163rd Infantry Division as well as a relief division that was numerically superior to the 163rd at the Battle of Suomussalmi. There are many reasons why the Soviets failed to defeat Finland, however the fact remains the Soviets sent untrained troops into winter operations with improper clothing and equipment. The Finns showed that a small force trained and experienced in cold weather operations could defeat a vastly numerically superior force. General Mannerheim, commander of the Finnish forces, estimated the dead at 200,000. He stated that most of these were wounded who froze to death because aid was not available. In his memoirs, Khrushchev placed the figure at 2 million.

1004. The German Invasion of Norway.

The German invasion of Norway was initiated on April 9th 1940, by a deception operation followed by sudden and decisive actions. Upon taking control of the ports and airfields, follow-on forces quickly responded with a full-scale invasion, transported by both ship and air, eventually building a force of 200,000 German troops. Because of Norway's terrain, the planning staff concluded they could in effect control the entire the country by executing landings at Norway's seven major cities: Oslo, Kristiansand, Arendal, Stravanger, Bergen, Trondheim, and Narvik. These population centers contain most of Norway's population, industry and trade. Additionally, to occupy these areas would result in the loss of half of Norway's sixteen regiments and most of the their artillery and airfields. For the invasion, Germany directed six divisions. "The complete destruction of the Norwegian Army was not considered possible as an immediate objective because of the size of the country and the difficulty of the terrain, but it was believed that the localities selected for the landings comprised the majority of the places which needed to be taken in order to prevent an effective mobilization and assembly of Norwegian forces and to control the country in general[1]." After the successful occupation of all the objectives, where only the occupation of Narvik was ever in doubt, military operations moved onto the Norwegian highlands. Because of the restrictive terrain, deep snow and steep valley slopes, German movement was restricted to the roads. The Norwegians established an effective defense by a series of roadblocks supported by fire from the ridges flanking the roads. Additionally, the German advance was delayed by the destruction of bridges along the roadways. The Germans developed a highly effective way to maintain momentum and contact by task organizing "... infantry spearheads organized in order of march as follows: one or two tanks, two truck carrying engineers and equipment, an infantry company with heavy weapons organized into assault detachments, a platoon of artillery, a relief company, relief engineers and artillery.[2]" As these units engaged the Norwegian defenses, units of ski troops would infiltrate to the Norwegian flanks. This resulted in the penetration of the Norwegian defense along several points. At Narvik, Norwegian and British forces isolated German troops. To bring relief, the Germans were forced to conduct a march on Narvik with 2,500 troops of the 2d Mountain Division through snow, rain and fog, without the aid of pack animals or vehicles; supplies to this force could only be delivered by air. Before this force could make it to Narvik, the allied forces withdrew. German military operations in Norway proved to be an outstanding success. Forcing the Allied withdrawal from Norway and occupying the major cities, Germany secured her northern flank, ensuring access to materials vital to the German wartime economy. Furthermore, Germany secured bases from which she could promote her submarine and air campaign against the British. For a more complete history on German operations in Scandinavia refer to Department of the Army Pamphlet No. 20-271, *The German Northern Theater of Operations*, 1940-1945, by Earl F. Ziemke, published in 1959.

1005. Operation Barbarossa.

On June 22, 1941 the German Army began the invasion of the Soviet Union, Operation Barbarossa. The plan, developed solely for summer operations, was delayed by more

[1] Ziemke, Earl F. *The German Northern Theater of Operations 1940-1945. Department of the Army Pamphlet 20-271.*

[2] Ziemke, Earl F. *The German Northern Theater of Operations 1940-1945. Department of the Army Pamphlet 20-271.*

than two months resulting from the Balkans campaign. Initially, the Germans achieved spectacular success, driving deep into Soviet territory. However, as time passed and German lines of communication (LOC) were stretched to their limit over poor highways, the weather situation changed. In the middle of September the autumn rains transformed the Russian plains and roads into a huge quagmire. With severely over extended LOCs, the Germans found it extremely difficult to move troops, supplies and equipment through the mud. On the night of 6-7 October the first snows began to fall. Completely unprepared for winter operations in the Russian hinterland, the German army was literally frozen in its tracks. Because Operation Barbarossa was executed in the summer no thought to winter preparations was given. More than lacking proper clothing equipment to protect the individual soldier from the elements, the German army lacked the knowledge to operate in a cold weather environment. Whereas the Soviets applied the lessons gained from their debacle in Finland, the Germans had to learn basics, such as how to keep aircraft operational in temperatures below freezing. More than this, German equipment failed because it was not designed to operate in a snow-covered environment. German tanks, with their narrow tracks, would sink in the snow whereas the T-34, with its wider tracks, maintained a greater degree of over the snow mobility. Colonel General Heinz Guderian said in November 1941, “This is sheer torture for the troops, and for our cause it is tragedy, for the enemy is gaining time, and in spite of all our plans we are being carried deeper into winter. It really makes me sad. The best intentions are wrecked by the weather. The unique opportunity to launch a really great offensive recedes further and further, and I doubt if it will ever recur. God alone knows how things will turn out. One must just hope and keep one’s spirits up, but at the moment it is a great test.” From December 1941 through March 1942 the Germans suffered a sickness rate of over 350,000 men, mostly resulting from frostbite.

1006. The Soviet Karelian Front Campaign (Northern Finland and Norway)

In 1944 the tide began to turn in favor of the Allies. Soviet campaigns were conducted against Germany on numerous fronts. In the spring the Soviets planned and conducted a successful campaign, which resulted in Finland suing for peace with the USSR and dissolving their alliance with Germany. At this point the German army hastily attempted to withdraw both personal (XIX Mountain Corps) and large quantities of supplies from the northern ports of Kirkens, Norway, and Petsamo, Finland. Soviet troops moved rapidly via the Lennigrad-Murmansk railroad in preparation for the Soviet invasion of the German occupied North Finland/North Norway. This combined arms campaign used all elements of modern amphibious warfare-advance force operations, ground combat elements, combat support, combat service support, and aviation including reconnaissance, interdiction and close air support. This operation must be considered a baseline for planning and conducting combined operations in the cold.

In the fall of 1944, some 56,000 German troops of the XIX Mountain Corps were occupying a strong point line just 70 kilometers northwest of Murmansk, about 200 miles north of the Arctic Circle. To clear these enemy forces from Soviet territory, Stavka ordered General K.A. Meretskov’s Karelian Front to plan and conduct and offensive, which was to be supported by Admiral A.G. Golovko’s Northern Fleet.

The Soviet force of approximately 96,000 men organized into a main attack force of two rifle corps, a corps-size economy-of-force formation and two enveloping forces. One consisted of two naval brigades and the other of two light rifle corps of two brigades each. On October 7, 1944, the Soviets began the offensive with a 97,000 round artillery preparation, followed by an infantry attack. They employed over 21,000 tubes of artillery and mortars, used 110 tanks and self-propelled guns, and enjoyed over-whelming air superiority. Engineer special purpose troops infiltrated up to 50 kilometers behind German forward positions to conduct reconnaissance before the battle.

While forces on the main axis where achieving a break-through against the German 2d Mountain Division, the light rifle corps were enveloping the German defensive position by a grueling march over extremely rugged terrain in the southern flank. Late on October 9th, a 3,000 man naval infantry brigade landed along the Barents Sea coastline behind the German northern flank, joined on the morning of 10 October by another naval infantry brigade attacking overland. Supported by naval special-purpose forces, the Northern Fleet conducted a second landing into the German left flank on October 12. Battered in the center and threatened on both flanks, the German corps executed a forced withdrawal into Norway. Soviet troops captured the port of Petsamo on October 15.

Soviet forces pursued the German troops westward, crossing into Norway on October 18 along the Northern flank and on October 23 on the southern flank. Naval infantry units assaulted German coastal positions on October 18, 23, and 25. Soviet troops captured a burning Kirkenes on October 25 and pursued the retreating Germans along the Norwegian coast as far as Tana Fjord, stopping October 30. In the south, the Soviets pursued the withdrawing German forces as far as Ivalo, Finland, stopping on second of November.

Soviet troops occupied the northern portion of Norway until October 1945. In November 1945, the Soviet forces withdrew, restoring the original Norwegian borders. For a more complete discussion of this battle, see FMFRP 12-24, *The Petsamo-Kirkenes Operation (7-30 October 1944)-A Joint and Combined Arms Operation on Arctic Terrain.* Also, Leavenworth Papers number 17, *The Petsamo-Kirkenes Operation: Soviet Breakthrough and Pursuit in the Arctic, October 1944.*

1007. The Battle of Attu

The Battle of Attu, May 1943, was a joint U.S. Navy and Army operation conducted in the northwest pacific against the Japanese. This extremely costly campaign was conducted in typically awful Aleutian Islands' weather and was marked by many costly errors. Approximately 12,000 troops sustained somewhat less than 3,000 casualties in 21 days of fighting. Intelligence was inadequate, terrain was not mapped or charted and often the fog was so intense that observation was impossible.

Planned rehearsal landings off the coast of Alaska were cancelled because of high seas and fog. The aviation combat element received only 5 days of training due to the late arrival of the carriers. The pre-D-day rehearsal landing was canceled due to foul weather. Amphibious shipping and troop equipment were short due to operational commitments in other theaters. Troop transports were overloaded with troops and cargo. D-day was postponed three times over a four-day period. Japanese defenders were at a high state of alert on D-day. The amphibious task force went north into the Bering Sea in the fog for three days. This resulted in the Japanese defense being relaxed at the time of invasion. The landing, conducted without air or naval gunfire support and under fog conditions, was accomplished in complete surprise, unobserved and without opposition. Landing beaches were widely separated by 30 miles. Waves of landing craft became lost in the fog on the way to the beach. Ships without radar became lost in the thick fog. Because of the fog, destroyers were unable to maneuver for fear of collision. Two collided.

Attu was defended by 2,300 Japanese troops who had been reduced to half rations because of the successful isolation of the area by the U.S. Navy. They established well dug-in positions in the high mountainous terrain. Scouts who parachuted onto the island in advance force operations reported the terrain un-trafficable for trucks and tracked vehicles. Movement of logistics severely limited combat power as foot troops were pressed into labor tasks. Movement of artillery pieces to firing positions was nearly impossible. The freezing cold, fog, bog-like tundra and steep, mountainous terrain proved harder to fight than the Japanese; 90 percent of the scout company and 75 percent of the reconnaissance troops suffered cold weather injuries.

1008. Marine Corps Experience in Korea

Marines are familiar with the experiences of the 7th Marines at the Chosin Reservoir in November 1950. The following comments are particularly indicative of what they faced:

LtCol Ray Davis, the battalion commander of 1/7, related of his battalion's performance on 10 November. He bedded down his battalion alongside a river in the Koto Plateau; the weather was so warm he bathed comfortably in the stream. Two evenings later the temperature was minus 16 degrees, with fierce wind. "When we got up in the morning," Davis said, "none of the vehicles would start. Troops had their noses turn white, big spots on them, and their fingers were numb. It was just an unbelievable change in the temperature in 24 hours." Colonel Alha Bowser watched Marines come off the line "just like zombies, the cold was so severe. Cold weather fighting is perhaps the most miserable type," he said. "There is nothing you can compare it with; wet, heat or anything. There is a sort of paralysis… at times that sets in, in extreme cold." When the savage weather struck, "Our men were not conditioned for it," concluded Colonel Homer Litzenburg, commander of the 7th Marines. "The doctors reported numerous cases where men came down to the sick bay suffering from what appeared to be shock. Some of them would come in crying; some were extremely nervous; and the doctors said it was simply the sudden shock of the terrific cold when they were not ready for it."

Not only did the cold have a devastating effect on personnel but also on operations.

"Everything was frozen. Plasma froze and the bottles broke. We couldn't use plasma because it wouldn't go into solution and the tubes would clog up with particles. We couldn't change dressings because we had to work with gloves on to keep our hands from freezing. We couldn't cut a man's clothes off to get to a wound because he would freeze to death. Actually, a man was often better off if we left him alone. Did you ever try to stuff a wounded man into a sleeping bag?"

Even decision making was affected:

"It was a numbing cold. I remember twice crawling down, my poncho over my head, with a flashlight, getting my map oriented to check out the direction. I would fix my hand over the maker, turn the light out and lift the poncho and get out to check the direction, and I wouldn't remember what had happened down there under the poncho. I'd get up and just stand there in a daze. Two or three people standing around would have a few words to say, and by that time I had forgotten what I was trying to do. I'd have to go down and do this thing all over again. Everybody had to repeat back to you two or three times to be sure of what was supposed to happen. We were just absolutely numb from the cold."

1009. The Falkland-Malvinas War, 1982

In April 1982, Great Britain responded to Argentina's invasion of the Falkland/Malvinas Islands. This action resulted from a dispute over the sovereignty of the islands between the countries of Argentina and the United Kingdom (UK) (see United Nations General Assembly Resolution 2056). Hastily mobilized, this amphibious operation resulted in victory for the UK in mid-June 1982. The Falklands, located in the Southern Hemisphere, are typical of arctic/sub-arctic regions and very much like the Aleutians, Greenland, Iceland and Scotland. They are wet, cold, with fjords, limited beaches, and have mountainous, tundra type terrain, limited villages and a low population density.

This operation required an amphibious deployment of over 8,000 miles. Over 8,000 UK troops were scattered all over the globe at the time of mobilization. Mount out was piecemeal; using nationalized civilian ships to augment the amphibious fleet. The British nuclear submarine force established sea-lane superiority early in the operation. In the Ascension Islands, ships rendezvoused, embarked equipment was sorted and reorganized, late arrivals married up their units and live firing of all weapon systems was conducted. Training was conducted en route including emphasis on physical fitness, training of personnel on call for fire procedures and medical training (self-aid, buddy-aid, and first aid).

Advance force operations began three weeks before D-day with the special air squadron/special boat squadron (SAS/SBS) conducting beach surveys, identifying naval gunfire missions and conducting a raid on Pebble Island. This raid destroyed Argentine turbo-propeller type aircraft (for counter-guerrilla operations/training) and turbo type aircraft (for training). **NOTE**: This did not affect air superiority in later actions. However, because the air strips at Port Stanley and Puerto Argentino were too short to accommodate the Argentine Mirage, A-4's, etc., the Argentines were required to use home based airfields some 400 miles away. The Argentines were not able to transfer expeditionary airfield components from Argentine bases due to logistical limitations. Before D-day, troops were crossed decked to amphibious assault ships and logistics landing ships. Whole blood was drawn from each man. Falkland ambient temperature of 40 degrees Fahrenheit/4 degrees Celsius preserved the blood.

The landing (D-day, delayed 1 day by foul weather) was conducted over one of the limited beaches in Sound Carlos Sound over 60 kilometers from Port Stanely. The British believed that the Argentines would defend against invasion from the sea. The British lacked an adequate distant early warning capability for their convoys, consequently they moved their logistical support base and air attack (Harriers) ashore (vice sea basing). A Rapier Battery (anti-air) was immediately established around the logistical support base. This battery combined with the hilly terrain and sea-based anti-air assets to afford protection form Argentine air attacks.

The Argentines established early success by attacking and sinking some ships. One, the Atlantic Conveyor, a ship-taken-up-from-trade, contained the helicopters intended for troop lift and numerous quantities of landing force supplies. The Argentine Air Force and naval aviation pilots were professionals and flew with great skill. Argentinean ground defenses were well prepared and mines were used effectively. However, they were often incorrectly oriented due to poor intelligence. Argentine troops were largely conscripts, many of them poorly trained. However, many fought skillfully and bravely. The dissension between officers and enlisted personnel was higher than normal, which could be expected of an inexperienced conscript force in wartime.

The British conducted deception operations including the naval gunfire shelling of numerous positions on both West Falkland/Gran Malvina and East Falkland/Soledad Island and SAS raids at Darwin and Goose Green. Because their support helicopters had previously been destroyed when the Argentines sunk the Atlantic Conveyor, British forces carried packs weighing over 110 pounds. The British attacked on two axes. The Royal Marine Commando and Army parachute battalions on the North Axis had to foot march over 60 kilometers over tundra, bogs and mountains to the final battle at Port Stanley. The Fifth Infantry Brigade, the Scots Guard, the Welsh Guard and the Gurkha moved over similar terrain on the South Axis. Logistical support provided by helicopters and BV-202s was limited and dedicated to the movement of artillery ammunition. Light tanks and BV-202s moved well over the terrain providing logistical and fire support. Antitank weapons were used to attack prepared positions. The Falklands taught that helicopters operating on or forward of the FEBA are at great risk due to the proliferation of handheld surface-to-air missiles. Consequently, the role of helicopters was reduced to logistical support.

The British, despite numerical inferiority and exceedingly long LOC, were determined, well trained and professional. Their units and staffs had worked together for years were fully confident in their abilities. Tactically, the British conducted all battles at night to gain maximum surprise and to conceal the inferior numbers of their attacking force. Squads consisted of two to three fire teams per squad with each fire team built around a machine gun. Attacks were conducted according to doctrine by using good intelligence collected through aggressive patrolling. Artillery fire supported each attack. They never attacked outside of their artillery fan.

The entire war was conducted in wet cold conditions in almost constant drizzle and rain. Troops were constantly wet. The cold, wet weather drained both forces. Discipline, professionalism and cold weather training in Norway paid dividends and contributed to the victory.

Trench foot was a significant problem among the British, who with heavy packs and constantly wet footwear, recognized that if the war had lasted longer, its influence would have undoubtedly been significant.

1010. Conclusion

Weather has proven a decisive factor in military operations. This is especially true of cold weather operations. History has shown us that the force trained to operate and fight in the cold weather will impose his will on the enemy. Forces inexperienced in cold weather operations fight not only the enemy, but the weather as well. During Operation Barbarossa, the period of December 1941 to March 1942, at times the German casualties from cold weather injuries exceeded those from enemy action. This is certainly a decisive factor, a factor that can be limited through training and education. Furthermore, history has shown that operating in a cold weather environment is not only an infantryman's concern; support agencies must be as adept at performing their mission under the same arduous conditions.

References:

Ambrose, Stephen E. *Citizen Soldiers*. New York; Touchstone, 1997.

Cooper, Matthew. *The German Army 1933-1945*. Lanham; Scarborough, 1978.

Dyke, Carl Van. *The Soviet Invasion of Finland 1939-40*. London; Frank Cass Publishers, 1997.

Engle, Eloise and Paananen, Lauri. *The Winter War*. Harrisburg: Stackpole Books, 1973.

Swinzow, George K. On Winter Warfare. Department of the Army Special Report 93-12.

U.S. Marine Corps. FMFM 7-21 Tactical Fundamentals for Cold Weather Warfighting. 1992

Ziemke, Earl F. (1959) The German Northern Theater of Operations 1940-1945. Department of the Army Pamphlet 20-271.

CHAPTER 2
PLANNING FOR COLD WEATHER OPERATIONS

2001. Cold Weather Considerations

Cold weather has a decisive impact on operations. It will affect personnel, weapons and equipment. History reveals that its greatest impact will be on the personnel operating in this environment. To be successful, Marines must be prepared to operate in adverse conditions. Leadership must ensure that Marines are prepared for the cold weather environment through environmental training. On the operational level, inclement weather will generally favor the defender. Attackers will be slowed by difficulties encountered by maintaining general mobility and proper logistics. On the small unit level, inclement weather will generally favor the attacker. Defending troops will be less alert and security will be difficult to maintain.

Cold Defined. There is no DOD definition of cold weather. The Marine Corps defines cold as:

Wet Cold	+40° to +20° F
Dry Cold	+20° to –5° F
Intense Cold	–5° to –25° F
Extreme Cold	–25° to –60° F

The Marine Corps definition of cold weather will be used throughout this publication.

1. Wet Cold. Wet snow and rain often accompany wet cold conditions, causing the ground to become slushy and muddy. Marines can become wet and cold if not properly equipped and prepared. Wet cold can occur at 60° F, particularly if there is a significant wind-chill. Wet cold leads to hypothermia and trench foot. Units must learn how to live and fight under wet cold conditions. Under wet cold conditions, temperatures:
    - Hover around freezing.
    - Stay generally above 20° F.
    - Alternate between freezing and thawing

2. Dry Cold. Dry cold conditions are easier to live in after the psychological shock of the cold has been conquered. Dry cold can be complicated by wind-chill. Under dry cold conditions:
    - Temperature ranges from +20° to –5° F.
    - Ground is frozen, snow is dry.
    - Humidity is extremely low.

3. Intense Cold. Intense cold exists from –5° to –25° F. It affects the mind as well as the body. Intense cold has a numbing effect. Simple tasks take longer and require personnel to use more effort than in temperate climates. Commanders must consider this when planning operations and giving orders for even routine tasks.

4. Extreme Cold. When temperatures fall below –25° F, the problem of survival becomes great. During extreme cold conditions, it is common for the individual to prioritize warmth and comfort above everything else. Personnel may withdraw within themselves and adopt a cocoon-like existence. This symptom may be reflected in group behavior. The commander must expect and plan for:
    - Material failures.
    - Vehicles to develop operational problems.
    - Munition failures.
    - Weapon malfunctions.
    - Heated shelter requirements for personnel to increase.
    - Increased level of supervision to keep Marines and equipment functional.

A. Seasonal Changes. Cold weather areas of operation are characterized by seasonal weather changes.

1. Winter. From the viewpoint of military science, winter can be characterized by a period when "troop and equipment movement is affected by snow and low temperature for at least one month per year." Using this definition, at least one-half of the world's dry land is in cold regions." (Swinzow, Special Report 93-12: On Winter Warfare) The severity of the winter season will change with the latitude and climate of the operating area. During winter conditions, commanders should expect:
    - A landscape that requires different clothing for warmth and camouflage.
    - Ground movement of men and machines to leave visible tracks.
    - Deep snow to slow down or completely stop off road fighting vehicles.
    - Water bodies to freeze and become possible roadways or MSRs.

2. Summer. The summer season typically offers a more temperate climate which is more hospitable to the Marines and equipment that operate in it. This does not hold true to the arctic regions of the world. The arctic has poor drainage and permafrost or muskeg will restrict mobility of operating forces. The sub-arctic, however, is not restricted by permafrost and muskeg. Once frost thaws and soil drains, cross-country movement can be accomplished.

3. Transitional Periods (Breakup / Freeze-up). Breakup and freeze-up are generally associated with spring and fall. In Marine Corps contingency areas, transition can and will occur almost instantly. This is due to the influence of maritime winds and warm water currents.

(a) Breakup. As river ice thaws, the surrounding countryside may flood, large ice jams may develop that will complicate problems. Traffic may be possible only at night when temperatures drop and ground surfaces re-freeze. Vehicles should carry reduced loads. Heavy traffic will turn unpaved roads into morasses. Movement may cease or become restricted.

(b) Freeze-up. Fall brings rains which complicate movement. Unpaved roads create deep mud. Ruts formed during the day may freeze at night, causing difficulties with torn tires, broken wheels and broken axles. As freeze-up progresses, the ground will become firmer improving cross-country movement.

2002. Command and Control

Marine units operating in the cold will often be isolated. Proper and timely warning orders will provide adequate time for subordinate commanders to prepare and respond to command direction. Mission-type orders are essential. Commanders must make every effort to keep their subordinates informed of their intent so those subordinate commanders can seize the initiative and feel confident in conducting independent actions. In order to maintain command and control in the cold weather environment, the commanders need constant, accurate information on the battlefield situation. Commanders must influence the battle by rapidly communicating direction and intent to subordinate units. Commanders must plan for the effects of cold weather on command and control so that it does not adversely affect the outcome on the battlefield.

A. Effects of the Cold on the Functioning of the Command Post. Cold weather will affect communication assets, personnel, mobility assets and security that allow the CP to function properly.

1. Communication Requirements. In a cold weather environment, the ability of the command to maintain effective communications will be diminished. Refer to Chapter 5002, Communication Considerations, for more specifics.

2. Personnel. Because the cold saps strength and energy, a three-watch system is recommended and should be planned for. Without augmentation, there are insufficient personnel to man the CP and displace command group elements. When task organizing for sustained operations in the cold, additional personnel must be planned for. Personnel should be sourced from parent units; for example, communicators and technicians from the communications battalion, watch officers from the major subordinate command headquarters, etc.

3. Mobility. Mobility assets provided to the unit may drive the CP configuration. If the unit is foot mobile and transports its equipment in team sleds, the CP will be austere. Ideally the CP will be more mobile. Two possible options are to use the typical HMMWV configurations or to use the BV-206. Both the HMMWV and the BV-206 allow rapid movement or displacement of the CP to preferred locations off the MSR where communications can be acquired and where security can be maintained. Units that are tasked organized to include LAVs and AAVs have the

ability to configure their CP around the command and control assets organic to these units (LAV and AAV C2 variants).

4. Security. Once the enemy locates a CP, it becomes extremely vulnerable. Cold and inclement weather will favor an infiltrating enemy's attack on the CP. For this reason, security is paramount and cannot be ignored. The larger the configuration, the more security is required. Concealment, remoteness, or inaccessibility may aid the CPs security plan. Regardless, an active defense of the CP is required.

B. Employment of the Command Post. Although command and control may become more difficult as units tend to be more widely dispersed and difficulties in communication increase, the general employment of the CP will not change in cold weather. If forces are to operate successfully, the CP must remain functional despite any difficulties associated with cold weather. Two elements of employing the CP in cold weather that take on added concern are displacing the CP and shelters.

1. Displacement. The commander must recognize that displacing the CP during cold weather or during inclement weather will take significantly longer. Displacement exercises are critical in a variety of environmental conditions to be successful on the battlefield. Future CP locations on the battlefield must be identified quickly and reconnoitered to ensure it will function effectively.

2. Shelters. The CP may be sheltered in a variety of configurations from tents and vehicles to buildings and expedient shelters. Regardless of its configuration, the CP must provide adequate warmth and shelter so the commander and staff can accomplish their mission. The method used to heat the CP must not compromise its location. Living spaces for personnel that comprise the CP or provide for its security should be located away from the CP. Possible CP configurations include –
   - A configuration of interconnected four-man tents.
   - General-purpose tents.
   - HMMWV configurations.
   - BV-206 configurations.
   - Existing buildings.
   - C2 variants of LAV and AAV's

C. Liaison in Joint or Combined Operations. The commander must constantly be aware of the need of liaison personnel when operating in joint or combined operations. Commanders must be prepared to provide knowledgeable subject matter experts to the host nation and other Services and to insist that the host nation or other Services provide the same to assist in the planning and execution of joint and combined operations. Host nation or joint support is an extremely valuable source of expertise. Liaison teams will ensure that this asset is used efficiently and correctly. Different communication links, capabilities, and SOP's may exist when operating with joint and combined forces. This can have a negative effect on operations if not considered when planning. Liaison staff personnel can help to solve this problem. It may also be necessary to provide appropriate communications systems to solve the problem.

These operating procedures and special equipment needs must be addressed and provided for during the planning process.

2003. Maneuver

Maneuver enables friendly forces to engage and destroy the enemy. This is vital, particularly in cold weather, but will take longer to achieve. The use of engineers to improve mobility, and equally, to prevent the use of critical terrain by the enemy will be a critical factor for the tactical commander. Sensible planning by commanders and their staff that take into account the capabilities of the individual Marine, his weapon systems and his level of training will prevent over-ambitious assumptions about movement in cold weather conditions. Mobility plays a major part in achieving successful maneuver against the enemy. The commander that makes positive use of the ground, air and suitable terrain to move troops and supplies will gain the tactical advantage that mobility provides.

A. Mobility. The speed of movement will depend not only upon the terrain and weather conditions, but also the training and equipment of the unit. Cross-country movement in cold and energy sapping conditions is difficult and time consuming. Keeping to trails, roads, ridgelines and valley floors are usually the easiest methods of movement, but it is also the most obvious to the enemy. Units using likely avenues of approach should exercise extreme caution.

B. Time and Distance. Estimating time and distance is an important military skill in cold weather. Time and distance factors will vary with the terrain, the climate and the season. Unusual weather and terrain conditions make problems of supply, medical evacuation, transportation, and services more difficult and time consuming. More time must be allowed for moving supplies and Marines because of the environment. Unless routes have been reconnoitered, precise time estimations cannot be confirmed. Attempts to increase speed can result in serious consequences such as loss of surprise, physical exhaustion and separating forces. Reducing navigational error is both energy and time saving.

2004. General Fire Support Considerations

Planning. Fire support planning in cold weather or mountainous terrain becomes more difficult because of concerns for terrain, weather and mobility. Refer to Sections *3001.B.3 Crew Served Weapons*; *3002 Artillery Operations*; and *Chapter 4 Air Operations* for more information.

A. Terrain. Terrain often will dictate what source of firepower will best support tactical operations. In steep terrain the mortar is highly effective. Artillery is generally more restrictive in steep terrain. Naval gunfire will be restricted by mountainous terrain. Targets in valleys or on reverse slopes will be difficult to effectively engage, due to the flat trajectory and high muzzle velocity of naval gunfire, even with reduced charges. Air support is extremely effective in all types of terrain, but will be less dependable during inclement weather or mountainous operations.

B. Weather. Current and accurate meteorological data will play a pivotal role in the Marine Corps' capability to provide accurate fire support. Seasonal conditions will dictate the most effective and proper mix of ammunition. Deep snow will dampen the impact of high explosive rounds and decrease the ability to use white phosphorus as a visible marker. This will increase the demand of variable-time and mechanical fuses as well as increase the demand of 81mm red phosphorus for marking purposes. Frozen ground will promote the use of high explosive rounds, since the fragmentary effects of the round will be augmented with frozen chunks of earth and ice.

C. Mobility. The mobility of fire support assets and the logistical capability to keep them re-supplied is critical to implementing effective fire support. Naval gunfire and air support are relatively unaffected by the mobility concerns prevalent to mortars and artillery.

1. Artillery. Snow, permafrost, swamps, bogs, and mountainous terrain limit off-road movement. The inherent limitations of the weight of artillery pieces limit movement in marginal terrain. The limited road networks and snow cover will complicate displacement of artillery batteries to new firing positions. Adverse weather and identifying signatures may limit displacement of batteries via helicopter lift.

2. Mortars. The current table of organization does not provide enough personnel in the mortar section to man pack (or ski) the mortar table of equipment and its associated ammunition into the fight over mountainous or snow-covered environment. If unit commanders are unable to provide mobility assets to the mortar section, the unit will lose its organic fire support

D. Applications of Fire Support. Although the principles of the application of firepower will not change from those in a temperate climate, our ability to accurately call for fire will become more difficult. Forward observers, forward air controllers, and Marines in general will be affected by –

Poor visibility due to snowstorms and increased fog in cold weather regions.
Unrecognizable ground features obscured by snow cover.
Difficulty adjusting fires caused by elevation differences between gun and target.

2005. Intelligence Considerations

Personnel in the intelligence community will face the same problems that other units will face in a cold weather environment. Intelligence personnel should be prepared to deal with the difficulties associated with basic survival, general mobility, equipment maintenance, and communications. Environmental training will be critical in providing personnel with the survival skills necessary. Most importantly the intelligence community must understand what affect the environment will have on both friendly and enemy forces. Determining the trafficability of routes and conducting terrain analysis for commanders will be critical to the commander in conducting successful operations. Refer to Section 5002: Communication Considerations and Section 5004: Motor Transport Operations to better understand the difficulties associated with communication and mobility.

2006. Logistical Considerations

The capacity of the Combat Service Support Element (CSSE) to provide adequate logistic support may be the determining factor when evaluating the feasibility of an operation. Commanders must always be prepared to alter the plan. They must be timely when issuing orders to allow adequate time for subordinates to react. This is even truer in the cold weather environment because of fewer passable roads and the volume of bulk stores required to support operating forces in harsh environmental conditions.

A. Classes of Supply. Chapter 5 discusses the difficulties of supplying conflict in the cold weather environment. Not only will the quantity of requirements increase (particularly classes I and III) but also the vulnerability of bulk stores, due to enemy identification of rear areas, will increase. The ability of the CSSE to move these stores will be decreased by limited road access, unless extensive snow removal equipment is used, and the inability of heavy haulers to move anywhere but on paved main supply routes (MSR).

B. Transportation. Operating forces will generally use the same organic assets in the cold weather environment that are available on various tables of equipment currently. The common misconception is that operating forces will be equipped with some sort of tactical over the snow mobility such as the Haaglund's Small Unit Support Vehicle (SUSV), BV-206 that units have trained with at the Marine Corps Mountain Warfare Training Center, Bridgeport, CA. The current inventory of this asset does not support a theater operation and that even if it did, the specific missions of unique assets (Tank, LAV, LVS, Engineering assets) cannot be met by the SUSV. Consequently, operating forces must learn the capabilities of organic assets in the cold weather environment and accompanying increased maintenance.

C. Engineering. All aspects of deliberate engineering, passive and active combat engineering are affected by the cold weather environment. Digging becomes difficult to impossible in frozen ground, heavy assets cannot move in loose powder snow, and class IV requirements will be limited, as all supply classes will be limited by road infrastructure and competing logistical demands.

D. Summary. This publication emphasizes that the environment, often more frequently than the enemy, is the critical vulnerability in surviving and fighting in the cold weather environment. This is never truer than for the logisticians. Comprehensive, realistic training in regards to support of the MAGTF in this environment is essential for success.

CHAPTER 3
GROUND COMBAT OPERATIONS

3001 Infantry Operations

Although the techniques and procedures applied by Marines in a cold weather environment may change, it is important to know that the tactics that have proven effective in a temperate, jungle, or desert environment will not. The basic principles of warfare remain the same. Two characteristics that should and need to be emphasized are leadership and discipline. History has shown that success on the winter battlefield comes from units that have demonstrated good leadership throughout the chain of command and had the discipline to conduct sound tactical operations in spite of the cold.

A. General Mobility Considerations. Successful operations will be dependent upon an operational force's mobility. The operating force that possesses greater off-road mobility will have an advantage on the winter battlefield. The commander needs to understand that mobility will vary as weather conditions change and that mobility may be measured in time rather than distance. The commander must make effective use of all available means of transportation and use each asset to its maximum advantage. The three assets that the infantry commander needs to be concerned with are foot mobility, vehicle mobility and air mobility.

   1. Foot Mobility. Movements must be carefully planned and executed with the understanding that distance can sometimes be as difficult to overcome as the enemy. During winter months, Marines can be confronted with conditions that range from bogs and permafrost to snow and ice. Snow conditions will vary greatly in depth and physical characteristics throughout the AO. Knowledge of current snow conditions in the AO will be essential to the commander during the planning process. Heavy snow cover will impede cross-country movement, hide terrain features, cover natural obstacles (such as brush, stumps and deadfall) and conceal minefields or other man-made obstacles. Non-compacted snow in excess of 30 cm (12 inches deep) will stop the movement of Marines on foot without snowshoes or skis. By training Marines in the skills required to use snowshoes and skis, this problem can be avoided. During winter months, snowshoes and skis are combat multipliers that enhance an operating force's mobility and open up vast expanses of terrain. This increases a unit's offensive capabilities, and forces the defender to expand his defensive perimeter. Skijoring (towing skiers behind a vehicle) is an advanced method of skiing that increases a small unit's mobility (Figure 3001-1 and Figure 3001-2).

TECHNIQUE	ADVANTAGES
Snowshoes	• Little training is required to gain a high degree of proficiency. • Little maintenance is required. • Carrying and pulling heavy loads on gentle terrain is relatively easy. • Movement in confined areas and around equipment is easy.
Skis	• Rate of movement is fast and efficient in terms of energy expended. • Carrying and pulling heavy loads on moderate to steep terrain is relatively easy (compared to snow shoeing).
Skijoring	• Each skier expends less energy if done properly. • Movement of a ski-borne unit can be expedited.

FIGURE 3001-1: ADVANTAGES OF OVER THE SNOW MOBILITY

TECHNIQUE	DISADVANTAGES
Snowshoes	• The rate of movement for a unit is extremely slow and inefficient in terms of energy expended. • Movement on moderate to steep slopes is difficult. • Movement through thick or cut-off brush is difficult.
Skis	• To become a proficient and efficient scout skier requires a lot of training. • Regular maintenance is required. • Movement in confined areas and around equipment is difficult.
Skijoring	• Skijoring conducted improperly can become tiring and slow. • There is an increased risk of suffering from a cold weather injury during skijoring operations.

FIGURE 3001-2: DISADVANTAGES OF OVER THE SNOW MOBILITY

2. Vehicle Mobility. The cold weather environment forces commanders to place an increased emphasis on map and terrain analysis. All possible routes should be designated throughout an AO. Reconnoiter all trouble spots or chokepoints by air or by foot patrols. Routes selected should take advantage of natural cover; provide concealment from air observation; and avoid steep slopes, abrupt ravines, unfrozen swamps, open streams, and other obstacles. Terrain at lower elevations often provides the easiest routes during winter months. Refer to *Section 5004: Motor Transport Operations.*

   a. Tracked Vehicles. Tracked vehicles may be entirely road bound during certain seasons (deep snow or spring thaw). Using them may depend on the availability and the security of these roads. Most tracked vehicles are slowed by a snow depth of 60 to 75 cm (24 to 30 inches). AAVs can increase the ground combat and logistical capabilities a great deal by crossing fjords, streams and rivers during certain seasons of the year. Tanks and AAV drivers will need extensive training and possible modifications to their track systems to successfully handle frozen surfaces.

b. Wheeled Vehicles. Motor transportation assets and light armored vehicles (LAVs) will be road-bound when the depth of noncompacted snow exceeds 75 cm (30 inches). Drivers will need extensive cold weather driver training, instruction in proper use of chains and self-recovery techniques. Studies of current contingency areas identify crowned roads with limited or no shoulders and deep drainage ditches alongside. These drainage ditches will most likely fill with snow and make it easier for vehicles to become stuck. Extensive snow clearing may be required to use road systems throughout the winter. It may be necessary to use host nation assets.

c. Over the Snow Vehicles. Over the snow vehicles (for example BV 206s and HMMWVs with a MATTRACK) are combat multipliers and the best alternative to off-road or on-road movement in cold weather operations. These vehicles have the capability to go off-road, however they are unable to negotiate extremely steep or icy slopes.

3. Helicopter Mobility. Helicopters increase mobility by providing logistical support as well as inserting and extracting forces away from MSRs and roadways. Helicopters shorten the time to move units and supplies. They are, however, subject to inclement weather, which is more prevalent during winter months. Aircraft lift capabilities will also appear limited due to the increased weight carried by each Marine and altitude restrictions.

B. General Considerations for Individual and Crew Served Weapons.

1. Planning for Cold Weather Effects on Weapons. When preparing to operate in cold weather it is important to understand the limitations of each weapon system. The infantry commander must ensure that his combat service support elements have planned properly to support increased breakages, cold weather lubrication concerns as well as a revised ammunition allocation to support cold weather operations.

a. There is an increase in the rate of breakage when weapons are fired at subfreezing temperatures. Firing pins, extractors, ejectors and small springs need to be carried at the small unit level, in order for spare parts to be on hand for immediate replacement. Extra mortar baseplates should be stocked in the battalion armory or available through the logistics train.

b. Lubrication of weapon systems is another concern when operating at cold temperatures. It is better to fire a weapon dry in the cold than it is to use an improper lubrication that will cause a malfunction during firing. Typically this is not a problem since, CLP does not freeze until it reaches –35° F. LAW (Lubricant, Arctic, Weapon) should be ordered for all weapons and used when temperatures range between 0° F to –65° F. History proves that other lubricants have been used cffcctively in extreme cold weather conditions when LAW was not available to fighting units. Some examples include graphite, kerosene and diesel fuel mix, and winter weight motor oil.

c. Effects of supporting arms against the terrain in a cold weather environment will cause commanders to adjust the typical allocation of ammunition. For example, frozen or ice-covered ground will increase fragmentation effect and deep snow can absorb up to 80% of fragmentation. There will typically be a higher demand for VT and mechanical fuses. Delay fuses can be used in a hard snowpack to help initiate avalanches or to break up frozen waterways. Red phosphorus is a better signaling device for CAS vice white phosphorus. Illumination on the deck is another effective signaling device in a snow-covered environment. Generally, more rounds will be needed to provide adequate indirect fire support. Plan suppression fires to accommodate slower rates of troop movements.

2. Individual Weapons Maintenance. Leaders will be required to ensure proper cleaning and maintenance is being conducted for weapons to be effective. This specifically needs to be done after unit movements or after large temperature changes. Marines should lightly lubricate weapons once cleaning is completed. Steps must be made to protect weapons and ammunition from the elements; however, they should be stored outside to avoid unnecessary condensation. To ensure weapon systems are capable of firing when required, SOPs must be established for the proper handling and carrying in snow-covered terrain. All points of entry into the weapon need to be covered to keep snow and ice out. If the muzzle is covered, it should be able to be fired through during immediate action. Any camouflage used on the weapon should not inhibit the functioning of the weapon system. Marines and their weapons need to be inspected regularly. Some examples of unit SOPs include: weapons slung, muzzle up in case of a fall; staging weapons and ammunition outside before entering a heated shelter; checking all covers – especially the ejection port cover – after a fall; and others. Refer to MCRP 3-35.1A Small Unit Leaders Guide to Cold Weather Operations, chapter 11 for detailed guidance to specific weapon systems.

3. Crew-Served Weapons. Mortars, machine guns, and anti-armor weapons have specific considerations. For more detailed guidance refer to MCRP 3-35.1A Small Unit Leaders Guide to Cold Weather Operations, Chapter 11.

a. Mortars. Mortars are employed in cold weather just as they are in any other environment. The challenge is establishing a gun line in snow or on top of frozen ground. If possible, mortars should be emplaced on unfrozen ground. If the ground is frozen, blast or break through the frozen crust with explosives or improvised materials to reach the earth. If this is not possible, create a platform from improvised material. Ensure aiming stakes are inspected often when placed in snow to ensure they are properly aligned. Ensure ammunition is protected from the elements and are at the same temperature as the guns. Ammunition should remain in its packing container until a fire mission has been called. Displacing and re-establishing a mortar position will take more time in cold weather. Plan for this.

b. Machine guns. Machine guns warm up very quickly. Breakage most often occurs in the early stages of firing. Gunners should warm up their guns prior to firing at the rapid rate. For example, gunners should fire short bursts prior to firing at the sustained rate. It is important to plan for 2 machine guns to cover one target. Since there is a higher degree of breakage, it is best to have a PDF or FPL covered by 2 weapon systems. If the ground is frozen or the snowpack is deep it will be time consuming to construct firing positions. Therefore, plan to construct more alternate positions, time permitting. Machine guns should not be moved over snow with an exposed belt. It is very easy for the weapon to malfunction if the belt is packed with snow. This will increase the reaction time of getting the machine gun into action. Gun drills should be conducted to lower this time.

c. Anti-armor. In a cold weather environment, the backblast areas need to be tripled to that of temperate environments. Positions selected should have minimal snow to better reduce the signature of the weapon system. If you cannot select an area with minimal snow, stamp the surrounding snow down with snowshoes. The Dragon, Predator, and Javelin are all man portable in a cold weather environment. The TOW will be restricted to roads and MSRs. Gunners should use appropriate UV eye protection when looking through a magnified sight over snow. Ensure gunners wear face, hands, and eye protection when firing due to the risk of burning launch motor propellants beyond the muzzle. The TOW, Javelin, Predator and Dragon are designed to fire down to -25° F.

4. Mobility of Crew Served Weapons. Distribution of limited over the snow vehicle assets to support crew-served weapons will be a critical factor on the cold weather battlefield. The HMMWV remains the prime mover for Weapons Company Platoons. Drivers should be trained in cold weather driving and be competent at putting chains on a vehicle. The introduction of the MATTRACK system to the HMMWV will greatly enhance the mobility of crew served weapons in snow-covered terrain. The BV-206s can also be utilized as a prime mover; some NATO countries have an 81mm mortar variant as well as a mount for the .50 cal. Snowmobiles should be effective at transporting Dragon, Predator, and Javelin teams quickly on the cold weather battlefield. This application can also be applied to machine guns and mortars. It is important to remember the limitations of the individual Marine. Consider augmented crew-served weapons teams for ammunition carrying purposes. It is well documented in S.L.A. Marshall's "The Soldier's Load and the Mobility of a Nation" that troops can only carry so much weight. This problem increases in winter, since each Marine is required to carry more personal gear than he does in a temperate environment. Each weapon system has the capability of being transported by man and sled.

5. Resupply. The logistical capabilities of a unit can and will dictate how much maneuver space a commander can use. Sound planning is required to anticipate

future needs. This is especially true before storms, when roads and MSRs may get shut down for days. The Weapons Company Platoon Commanders and Platoon Sergeants need to be aggressive in their planning and coordination with their logistical support.

6. Optics. It is important to understand the capabilities of optics in the cold weather environment and the effects that snow will have on their operability.

    a. Light amplification systems work better in snow-covered terrain due to ambient light refracting off of snow. Nights are longer in winter. They become even longer as forces move toward the North or South Pole. Night vision devices are a great asset in the winter. Remember that batteries will drain faster in the cold and that they will be used more due to longer nights.

    b. Thermal imaging systems work better in the cold due to increased contrast between the cold background and the warm target. It is difficult, but not impossible, to utilize thermal camouflage to defeat these systems. Some methods include building snow walls, placing one foot of snow on a poncho as overhead cover, wetting surfaces, and using reflective insulation. Building heat-generating devices for deceptive purposes is another method to confuse the enemy. Understand that just walking or skiing through snow will leave a thermal track that can be seen for a short period of time. Thermal systems are a great asset in winter and should be used. Understand that counter-measures may be used.

    c. Laser target designators function properly in cold weather. However, it has been documented that snow and ice can cause a laser beam to bounce off the target and illuminate a point miles in the distance. This would cause the bomb or missile to track the wrong signal.

C. Scout Skier Employment. Ideally an infantry battalion operating in a snow-covered environment will have Marines that are specifically trained as scout skiers. A scout skier is a Marine that has been given dedicated training to enhance his ski mobility and is capable of operating independently for extended periods with limited logistical support. Each line and weapons company should have a scout skier platoon. Refer to Figure 3001-3 and Figure 3001-4 for an example of a line company and weapons company task organized scout skier platoon. The battalion commander has the prerogative to integrate the platoons into a scout skier company or to leave the platoons as a company level asset. Scout skiers will enhance a unit's combat effectiveness during movement, offensive operations, and defensive operations.

LINE COMPANY SCOUT SKIER PLATOON	
1	Platoon Commander

	- Capable of making sound tactical decisions - Capable of taking mission type orders and carrying out commander's intent - Competent in using supporting arms assets - Experienced in cold weather (Winter Mountain Leader)
1	Platoon Sergeant - Capable of assuming role and duties of platoon commander - Capable of independently requesting all necessary logistical supplies
1	Radio Operator - Capable of maintaining and providing communication during cold weather operations
3	Squads - each squad consist of approximately 8 men - each squad have 2 team tents and sleds
ALL MEMBERS OF THE PLATOON MUST POSSES ENHANCED SKI MOBILITY	

FIGURE 3001-3: T/O SCOUT SKIER PLATOON – LINE COMPANY

WEAPONS COMPANY SCOUT SKIER PLATOON	
1	Platoon Commander - capable of making sound tactical decisions - capable of taking mission type orders and carrying out commander's intent - competent in using supporting arms assets - experienced in cold weather (Winter Mountain Leader)
1	Platoon Sergeant - capable of assuming role and duties of platoon commander - capable of independently requesting all necessary logistical supplies
1	Radio Operator - capable of maintaining and providing communication during cold weather operations
1	Machine Gun Squad - consist of approximately 8 men - have 2 team tents and sleds
1	Mortar Squad - consist of approximately 8 men - have 2 team tents and sleds
1	Anti Armor Squad - consist of approximately 8 men - have 2 team tents and sleds
ALL MEMBERS OF THE PLATOON MUST POSSES ENHANCED SKI MOBILITY	

FIGURE 3001-4: T/O SCOUT SKIER PLATOON – WEAPONS COMPANY

1. Scout Skiers During Movement. Scout skiers can be used during movement to contact as the lead element. In this manner they can navigate and break trail for the company that is following in trace on snowshoes. Scout skiers can also use their increased mobility to provide flank security for a unit during movement. The

commander may also desire to use his highly mobile scout skiers as the company reserve in order to use them as an enveloping force once contact has been made.

2. Scout Skiers in the Offense. Scout skiers can be used in the attack as an enveloping force or can be used to exploit success as the company reserve. Scout skiers also possess the capability to conduct long range extended patrols for the commander.

3. Scout Skiers in the Defense. Scout skiers provide the commander with a quick and mobile counterattack force in the defense. They also provide the capability to provide the commander with long-range patrols in front of his defensive position.

D. Offensive Operations. Superiority in combat will go to the force that is trained to live and move in the AO and is less restricted and reliant on road networks. Trained units will be able to expand their space of maneuver by moving away from roads and attacking their enemy's weaknesses. Untrained units will become road-bound due to the increased CSS requirements. Marines must resist the natural temptation to stay close to the lines of communications that provide access to this support. Over the snow movement is demanding and time-consuming, but critical for successful operations.

1. Aims of Offensive Operations. The tactical aim of offensive operations in cold weather is the same as in any other military action: *destroy the enemy force and his will to fight*. This can be done in a number of ways but consideration to the following factors will assist the commander in achieving his aim.

   a. Forces of Nature. A unit that is able to use the elements and terrain to their advantage will be successful. Training in cold weather during adverse conditions will increase a unit's survivability and mobility. In numerous campaigns and battles fought in cold climates (see Chapter 1), the cold has been a significant factor in producing casualties. Troops who have been exposed to the cold and inclement weather during training exercises will be better prepared in a combat situation. The commander that understands the limitations that the cold, wind, rain, snow, ice and mud will place on his forces and understands how these elements will affect his enemy will best be able to use the forces of nature to its fullest potential.

   b. Use Deception Effectively. By employing the tactic of frequently moving and shifting positions, the friendly force commander can cause the enemy to shift his disposition of forces. This will require unnecessary energy expenditure and an increased burden on the enemy's logistical train. Friendly troops, if prepared, can take advantage of this. Additionally, track and thermal deception methods can be employed to ensure that enemy forces have difficulty maintaining the accurate location of friendly troops.

   c. Attack the Enemy's Command, Control, and Communications. Command, control, and communications (C^3) in the cold are difficult to establish, operate, maintain, and displace. In large operational areas where positions are often over-stretched, C^3 sites become increasingly vulnerable. Flanks and rear areas are often left lightly defended and present excellent opportunities to infiltrate and offensively target C^3 nodes.

d. Supporting Arms. Supporting arms are as important in cold weather as they are in any other environment. Supporting arms can be sudden, violent, and decisive or continual and harassing, robbing the enemy force of sleep and rest. Accurate reporting and fire mission requests are critical to ensuring that valuable ammunition is used effectively. The smothering effect of snow on explosives and the construction of the enemy's defense must be considered.

e. Maintain Constant Pressure on the Enemy. The tactic of keeping the enemy in a constant state of alert in order to weaken his morale and will to fight is extremely effective in the cold. Commanders should not allow their offensive action to slow to the point that enemy forces can recuperate and possibly seize the opportunity to take the offensive. The freedom of the enemy's maneuver should not be limited by their ability to cope with the climate and terrain but by the offensive actions of friendly forces.

f. Develop Psychological Advantage. A psychological advantage is often gained as a result of successfully conducting offensive operations. By maintaining the initiative, friendly troops remain active and alert (provided that they are sufficiently rested, fed, and hydrated). Conversely, constant incoming fires and the preoccupation of coping with the environment can demoralize enemy troops in the defense. This will weaken their resolve to fight and lead them to become lethargic, withdrawn and less alert.

2. Attacking an Organized Position. Tactics remain the same as in temperate zones. The mission and commander's intent must be provided as early as possible to provide direction for the staff and subordinate commanders. The attack order is formulated using the standard 5-paragraph order, with an emphasis placed on keeping personnel informed. Particular emphasis should be given to planning for logistical support that will require additional time to plan, process, and execute.

a. Main and Supporting Attacks. The opportunity for maneuver is usually present in cold weather operations. Main attacks usually are directed against the flanks or rear areas, while supporting attacks are directed against the enemy front to fix the enemy in place. Additional forces may be employed to bypass the enemy position and cut enemy routes of reinforcement or withdrawal. The most mobile troops should be the main effort or used to reinforce or exploit success.

b. Control Measures. Control measures should be used as they are in temperate zones. Avoid using streams, gullies, or roads (easily identifiable on maps) that may be covered by drifting snow. An increased reliance on azimuths may be required if terrain is barren, flat, and snow swept.

c. Supporting Arms. Prepared fires should be closely coordinated. If possible, forward observers should be included in reconnaissance patrols. Avoid displacing organic fire support assets during the attack, since preparing firing positions is time-consuming and logistical support is difficult.

d. Engineering. Engineering materials and bridging equipment must be anticipated. Move engineer support as far forward as possible. Coordinate and use host nation assets to the fullest.

e. Communications Planning. The communications plan is made in detail and must provide measures (built in redundancy) for overcoming difficulties peculiar to the environment.

f. Logistics. Supply reserves must be kept mobile and well forward (particularly classes III and V). It will be necessary to establish rapid re-supply points in forward areas.

3. Preparing for the Attack. Preparations for the attack should mirror preparations in a temperate climate. Concurrent activity is a key element during this phase of operations. With Marines preparing weapons, skis, and snowshoes; conducting rehearsals; as well as feeding, hydrating and resting; an increased emphasis must be placed on detailed orders and the timely use of warning orders. The commander should pay special consideration to the commander's reconnaissance, trail-breaking, movement to contact, and actions in the assault position.

a. Commander's Reconnaissance. Due to the problems associated with tracks being left in the snow, the commander may have to delay his reconnaissance until the last possible moment. He may have to rely on his reconnaissance assets relaying information back to his position, where he will then depart for offensive actions with his unit. An alternative is for the commander to conduct his reconnaissance with security, sending guides back to the main body to lead them into position. This form of reconnaissance should be used to confirm the plan with only minor changes being briefed to subordinate commanders prior to commencing into the attack.

b. Trail-Breaking. The additional task of breaking trail from the assembly area to the assault position must be included in orders with ample time and personnel to complete the arduous task. Time and personnel requirements will have to be accurate to ensure the trail breaking unit is not left in the assault position too long or the main body is not held up by a slow moving trail breaking party. Training and accurate map analysis will ensure timings are precise. Consideration must be made to ensure the trail breaking group has not expended so much energy that they are too exhausted to conduct their role in the attack. Another consideration for reconnaissance is that the reconnaissance unit will leave a trail in the snow behind them.

c. Movement to Contact. Normal movement drills must be maintained with consideration to the extra difficulties of moving through snow with sleds in cold temperatures. Troops must be allowed to take adequate breaks and sled hauling responsibilities must be rotated amongst their teams. Track discipline must be adhered to not only for security but also to ensure the assault troops do not have to break trail in fresh snow. All this is done to ensure Marines do not arrive at the assault position exhausted.

d. Actions in the Assault Position. Actions should be conducted in a smooth manner and should be rehearsed prior to the attack. Once forces reach the assault position, they halt long enough to make final preparations for the attack. Security should be paramount. Depending on the method of mobility that will be used for the attack, skis and snowshoes will be staged in a centralized location or drug behind assaulting troops. Packs and sleds should be staged and Marines should carry only the equipment required for the attack. If weather is extremely severe, warming tents should be constructed for re-warming and casualty collection purposes as required. Supporting weapons and ammunition should be moved forward and staged at selected firing positions on team sleds. A simple signal should be used to begin the next phase of the attack. It is important to emphasize that all actions should be swift and deliberate in the assault position to avoid any unnecessary delay. The assault position should be located as close as possible to the objective without becoming unnecessarily vulnerable.

4. Conduct of the Attack. If possible the attack should be conducted from high to low ground with the wind at the back of the attacking force, forcing the enemy to face into the wind. While movement to the assault position may be on skis, the attack should be on foot or snowshoes. Snowstorms or inclement weather will aid an attacking force. Keep formations compact until the engagement begins. Plans should be kept simple. Have limited objectives and have a positive means of identification (especially when both combatants are wearing overwhites). The tempo of the attack should be such that troops do not remain exposed to the elements for prolonged periods of time.

5. Consolidation. Normal re-organization drills should take place once the objective has been assaulted. Medical evacuations and the re-warming of cold, wet Marines must be addressed quickly but kept in perspective of accomplishing the mission. Re-warming tents should be constructed and located in the nearest available concealed area and brought onto the objective immediately following attack. Begin re-warming Marines as soon as possible without sacrificing security. Relieving assault elements immediately with reserve elements is ideal and should be conducted when possible.

6. Exploitation. Caution must be used when exploiting success in the cold weather environment. While exploiting success can allow the commander to attack the enemy's critical vulnerabilities, the detrimental effects of maintaining the tempo of the attack at the expense of cold, wet and tired assault elements can be deadly. One

method for a commander to exploit success without further committing friendly forces is to use supporting arms assets. If the commander chooses to pursue the enemy, the forces chosen should have high degree of mobility. They can be mounted on skis, over the snow vehicles, or helicopters. Ideally the troops chosen will be well rested. Great rewards are available to the commander that makes preparations to exploit success prior to commencing in the attack.

7. Medical Evacuation. The momentum of the battle should not be slowed down for casualties. However during consolidation, commanders must be prepared to assist in the medevac of the injured. Deep snow and cold temperatures will adversely affect the ability to medevac casualties quickly. Casualty handling teams, sleds, and over the snow vehicles must be identified for moving casualties to collection points. This effort must be rehearsed in exercises and included in the initial formal order. Planning must include using organic tents and vehicles as unit aid stations.

8. Security in the Offense. Security requirements during offensive operations in cold weather are no different from those in temperate zones. The only difference is the amount of effort that is required to conduct proper security operations due to the weather.

9. Tactics and Techniques for Cold Weather Offensive Operations. While the tactics used to attack and destroy the enemy are the same as they are in temperate climates. Some tactics and techniques have been proven to be extremely effective during cold weather operations. Two good examples are the infiltration techniques applied by the CCF in Korea as well as the Motti Tactics developed by the Finns to counter the Soviet Offensive during the Finnish-Soviet War from 1939-40. Deliberate ambushes are another effective tactic that can be employed in cold weather.

a. Infiltration is an effective method of massing units for specific operations. The difficult consideration about infiltration is providing for the needs and re-supply of these units. Infiltrating troops generally must carry all necessary weapons and ammunition to conduct the attack. Additionally, they must carry food and the essentials for survival. There are four situations conducive to infiltration. Each of these situations is more prevalent in cold weather environments than in temperate climates. They are fatigue, difficult or seemingly impassable terrain, periods of reduced visibility, and enemy position.

(1) Fatigue. The harsh environment takes a toll on the human body, which burns calories in an attempt to generate enough heat to keep the body warm. Extreme temperatures and high winds sap strength, lower morale, and cause units on the defense to be less alert.

(2) Difficult or Seemingly Impassable Terrain. History shows many examples of forces that have relied on seemingly impassable terrain on which to anchor a flank / rear. Often offensive movement over this terrain leads to surprise and defeat for the defending force.

(3) Periods of Reduced Visibility. Snow, rain, fog, and long winter nights provide periods of reduced visibility that increase susceptibility. Modern weather predicting capabilities enhance the Marine Corps ability to exploit these periods.

(4) Enemy Dispositions. In remote winter areas, forces are likely to be small in relation to the vastness of the areas with which they operate. Units in defensive positions will probably be covering more ground than in normal climates and therefore, more susceptible to infiltration.

b. Motti Tactics. The Finnish word motti refers to a pile of logs ready to be sawed into lumber. The Finns developed motti tactics to combat the Russian offensive in 1939-40. They were most successful in the forested, snow-covered area of Finland. The Finns understood that the Russian forces were not prepared for winter warfighting and almost totally bound to the roads, with the exception of just a few ski troops. Motti tactics are important to understand as Marines since we may opt to use them in a cold weather environment or face an enemy attempting to use them against us. Certain conditions must be to successfully employ motti tactics.

(1) The attacking unit must -

- Have superior off road mobility.
- Be able to live in the attack zone to preserve the continuity of the battle, despite snow and cold weather.
- Be able to navigate at times of decreased visibility (darkness, storms, fog, and dense forest) while maintaining good cover and concealment.
- Maintain the element of surprise.

(2) The defending unit must –

- Be almost totally road bound.
- Have poor off road mobility or be reluctant to conduct off road reconnaissance.

(3) The defender must be forced to stop in order to present a linear target to the attacking force. On the battlefield this is accomplished as deep snow, mud, tundra, steep mountainous terrain, fjords, or a combination of these prerequisite conditions fix the road bound enemy. By use of deliberate engineering (avalanches, blown roads, or bridges), the enemy can be stopped to present a linear target.

(4) Motti tactics are generally conducted in three phases. The GCE commander controls the overall operation and establishes the zones of activity for the battalions. He also retains indirect control over attacking units of company size or larger.

(a) Phase One – Locate and Fix the Enemy. The initial reconnaissance locates and fix the enemy force. Following the initial reconnaissance, fighting patrols swarm on the enemy from all directions. These patrols vary from a squad to a platoon in strength. Ambushes and hit and run tactics are used to –

- Disturb the composition of the enemy.
- Create an air of uncertainty.
- Prevent uninterrupted sleep and rest.

Consequently, the enemy is compelled to use more forces on security tasks. These patrols create the illusion that attacking forces are everywhere and the enemy never knows where to expect the next attack. Maximum use is made of the cover of darkness, terrain, and concealment from forests. Enemy security is avoided. Patrols execute demolitions and plant mines in the rear of the enemy. Patrols carry out more than one task. Sporadic attacks in company or battalion strength are made directed at specific limited objectives. After the objective is destroyed or the enemy is forced to deploy, the attacking force disengages.

(b) Phase Two – Attacking and Cutting. In this phase, surprise flanking movements and envelopments are carried out. This phase may be subdivided into three consecutive activities:

- Movement into assembly areas.
- Movements into attack areas near the objectives.
- Enveloping and cutting.

By these maneuvers the enemy column is isolated and then sliced into small groups, each of which in turn is isolated.

(c) Phase Three – Isolation and Annihilation. As the enemy exhausts himself in efforts to break out, the main force regroups and repeats its cutting phase. The isolated motti are spit gain by attacks on the flanks. Enemy LOC must be kept closed. Units are dependent upon their LOC for survivability; without fuel and food, troops will die in a cold weather environment. If there are built up fortifications, some softening will be necessary by sniper, direct fire weapons, harassing fire, and propaganda. The commander must always consider supporting arms (Artillery and CAS) when conducting motti tactics.

c. Deliberate Ambushes. Snow covered terrain offers favorable conditions to mount deliberate ambush operations. The terrain will channel the enemy (especially one with poor mobility) into valleys and depressions as well as onto roads. Deliberate engineering tasks can hinder the advance of the enemy by blowing bridges and roads as well as causing avalanches to block routes. Attacking forces must battle the elements while waiting to spring an ambush. Time and consideration must be made to prepare positions that offer some comfort and protection from the elements. Constructing warming shelters in a covered and concealed position is recommended to keep Marines alert in the ambush position.

E. Defensive Operations. The principles of the defense do not differ in cold weather operations from any other environment. Certain aspects of the defense should be considered with greater emphasis due to the unique characteristics that the environment places on the individual Marine. These include:

- Conduct the defense aggressively. Keep Marines active. This will not only keep them alert, but warm as well. Maintain activity by conducting frequent patrols, rehearsing counterattack procedures and observing stand-to procedures.
- Take advantage of high ground when feasible. Force the enemy to attack uphill, which is extremely difficult in deep snow. Armored vehicles will have difficulty climbing even moderate slopes when the ground is frozen.
- Use natural obstacles. These may be created by the weather or be features that exist throughout the year. Engineer initiative can be a force multiplier in the cold weather environment.
- Use prevailing winds effectively, if possible. Force the enemy to attack into the wind. This will force the enemy to fight two battles, one against the weather and the other against the enemy.
- Use prepared defensive positions. They provide great advantage to the defender for both survival from the enemy and the environment.
- Develop self-supporting defensive positions. A commander must maintain his own security and not rely on supporting forces.
- Develop a comprehensive barrier plan. Linking this facet with security patrols and a good communication plan will enhance the defense of the position.

1. Assuming the Defense. Since terrain and weather will restrict the mobility of the attacking enemy, the defender has numerous advantages. The defending commander must ensure that he maintains an active defense and conceals his position until his fires can be used with maximum effect. Marines must be prepared to fight the defensive battle in any weather conditions.

2. Weapons. The commander should employ his weapons as he would in any environment. Refer to Section 3001.B: General Considerations for Individual and Crew Served Weapons. Some points to remember:

    - Machine Guns. The machine gun will offer the commander a primary source of firepower from the defense. Snow and cold should not be allowed to diminish machine gun performance. Time must be allocated to mount all machine guns on stable firing platforms. Marines must practice this technique in a variety of snow and ice conditions in order to be effective in all conditions. The practice of freezing tripod / bipod legs into permanent position is not recommended, because it limits your capability to displace the weapon system. In extreme cold temperatures ice fog will form around the trajectory of the round. This will pinpoint the location of the machine gun and gun crew to enemy attackers. This will increase the use of alternate positions in a cold weather environment.

- Mortars and artillery lose much of their effectiveness due to the smothering effect of deep snow. Mechanical and VT fuses should be used to counter the smothering effect of snow. If ground is frozen and without snow cover, HE rounds will be more effective against the enemy.
- Engineering efforts, properly organized, can be significant weapons on the cold weather battlefield. Commanders should employ their skills to the maximum.

3. Terrain Selection. Considerations when selecting ground should be no different than the terrain one selects in a temperate environment. One additional consideration is to ensure adequate drainage during seasonal changes. Seasonal changes can cause fighting positions to fill with water. Fighting out of positions filled with cold water can be demoralizing and cause unnecessary environmental casualties from hypothermia to trenchfoot.

4. Obstacle Construction. Construction is often limited by the frozen ground surfaces. Obstacles can be constructed of snow, ice, or snow and ice, and can be frozen in place (refer to *Appendix B: Fieldworks and Camouflage*). The effect of these obstacles is subject to the continued cold weather. Warming temperatures will melt snow and ice obstacles and will often produce natural obstacles like flooding or ice slush. Areas in the defense where there is little snow or which is easily traversed by the enemy should be reinforced with artificial obstacles. Use wire entanglements (especially concertina), pitfalls, abatis, antitank mines, and antipersonnel mines. Cover these areas with fire. The enemy's use of frozen waterways can be denied by laying minefields in the ice.

5. Deception. Deception measures must be taken to add to the overall strength of the position. The enemy must believe the defending troops are either in an alternate location or underestimate the true strength of friendly forces occupying the defense. The commander can accomplish this by employing trail deception, dummy positions, camouflage and false helicopter inserts.

   a. Trail Deception. By laying deceptive tails in the snow the enemy will not be led directly to the defensive position. If done properly, the enemy will be led to an area covered by defensive fire and observation, preferably the defense's kill zone. Additionally, patrolling forces from the defensive position must incorporate measures that will confuse the enemy as they attempt to analyze friendly tracks. These measures will ensure the enemy cannot determine the full strength of the defending force.

   b. Dummy Positions. A dummy position should be established within the defending troops field of observation and fire. The dummy position should be realistic in its construction and size. Additional measures can be taken to occupy the position temporarily. Lighting fires can provide enemy forces with false thermal signatures.

c. Camouflage. Good use of camouflage within the defensive position should play a major part in the defensive plan. This measure will confuse the enemy's efforts in determining the exact location of all forces as well as the actual size of the defending force. As in any other environment, camouflage should be continuous. Movement within the defensive position should be minimal and carried out by troops in full camouflage.

d. Helicopters. The use of false LZ's also add to the confusion of the enemy. The use of helicopters in the cold weather environment in addressed in Chapter 4, Air Operations.

6. Conduct of the Defense. The commander should conduct the defense as he would in any other environment. Pay special attention to individual Marine preparedness. Weapons must be properly maintained for immediate use. Also inspect the general health and hygiene of Marines regularly. This will help prevent a high number of cold weather casualties.

7. Defensive Positions. There can be no substitute for traditional defensive emplacements; however, in cold conditions the depth of permanently frozen ground may prevent the construction of these positions. Normal entrenching tools are not sufficient to penetrate the frozen ground and the use of snow shovels is not recommended. If engineering support is available or troops have sufficient knowledge in demolitions the use of explosives has proved successful. The following is an extract from an actual report covering the defense of Hagaru in Korea in 1950:

> "How and Item Companies were ready. All platoons were well dug-in, though the earth was frozen to a depth of 6 to 10 inches. The men of Item Company used their heads as well as their hands after Lt. Fisher managed to obtain a thousand sandbags and several bags of C-3. The explosive was placed in ration cans to make improvised shape charges that blasted holes through the frozen crust of snow and earth. Then it became a simple matter to enlarge the holes and place the loose dirt into sandbags to form parapets. This ingenious system resulted in deluxe foxholes and mortar emplacements."

Snow defenses are subject to the weather and will break down under sustained fire. They may often be the only option for troops in areas of deep snow. Snow combined with other materials will provide adequate protection if used correctly and in sufficient depths. Commanders must ensure their troops know what level of protection is required to offer protection against enemy arms.

a. Materials. The amount of natural materials required to construct a full defensive position is vast and produces a huge logistical problem. Commanders must ensure their immediate area is not stripped of trees being used for barriers and reinforcements. These trees provide essential camouflage and must be left in place. Materials will need to be transported to the defensive position. Used ration boxes can be implemented into the defensive positions to strengthen the fortifications. A recent trend in defensive measures has been to use bales of hay or straw soaked in water and allowed to freeze. Not only does this provide a strong material, but also it offers a shaped block that is easy to build with. Farms holding large supplies of baled materials can be approached and their materials utilized. Measurements of levels of protection offered are currently available. Refer to *Appendix B: Fieldworks and Camouflage.*

b. Camouflage. Once construction is complete the defensive commander may need to address the problem of camouflage of his area again. With the amount of movement over and around the position tracks will have left signs of the location. These need to be concealed.

c. Time. The time taken from the initial reconnaissance of the proposed position to the establishment and construction of the defenses can be a long affair. The defensive phase must be planned to justify the extent of the work involved in a deliberate position. The practice of moving positions often becomes a tiresome drain on manpower and resources.

8. Composition and Location of the Reserves. An aggressive defense will require the formation of a proportionately large reserve. The reserve should have a high level of off-road mobility. This mobility can be individual (skis or snowshoes), wheeled or tracked vehicle or helicopter. If the commander is relying on individual mobility, it is recommended that ski trained forces occupy the reserve. Regardless, the reserve should be placed in a covered and concealed position, protected from enemy supporting arms. Trails and roads to the probable points of action should be prepared and camouflaged.

9. Preparing for the Counterattack. Actions for a counterattack need to be fully briefed and known by all. The counterattack should ideally take place downhill with rally points on the reverse slope, if possible. If conditions on the battlefield allow, the whole counterattack should be rehearsed. These actions should be practiced in slow time during daylight hours until the counter attack force is capable of conducting the attack full speed during the hours of limited visibility. Commanders should make a thorough appreciation of the value of rehearsals against the potential for compromise of the plan by tracks left in the snow or by enemy observation. Use the elements (for example a large snowstorm) to cover up tracks after a rehearsal. Timing is everything. Drills or SOPs for counterattack should include considerations for the following:

- Pre-established rally points.
- Identified routes of attack.
- Caches of ammunition.
- Developed crew served weapons positions.
- Identified firing positions that can destroy previously established crew served weapons.

3002. Artillery Operations

A. Appreciation of the Cold.

Employment of artillery in cold weather operations requires imagination, detailed planning, training, flexibility, initiative, and effective SOPs. Wet cold, extreme cold, deep snow, and difficult terrain create many problems not encountered elsewhere. Cold weather changes time and distance factors. Experience has shown that it can take up to five times normal to perform even simple tasks.

Artillery commanders must understand the effects of cold weather on their personnel and equipment. All aspects are affected. The cold increases equipment maintenance and material requirements, ancillary equipment requirements; and creates requirements for additional equipment. Since virtually every task takes longer, how to move, when to move, and how far to move become critical considerations.

Positions in snow-covered terrain are vulnerable to attack from any direction. Snow and limited MSRs may make it impossible to evacuate guns. Crew must be trained in demolitions and destruction of field pieces and munitions. Artillery units must take maximum advantage of peacetime cold weather training to have the greatest possible number of experienced artillerymen available for combat in cold weather operations.

B. Tactical Considerations

1. Planning. The artillery commander and his staff must closely participate in the supported command's planning process, from commander's guidance through the issuance of the final order. Such participation allows the artillery commander to plan for proper positioning and support requirements and makes the maneuver commander aware of limitations placed on artillery by the cold. The artillery commander can then properly plan, position, fire, and resupply his firing elements.

2. Positioning. Positioning must support the expanded battlefield inherent to cold weather operations. Supply lines will be lengthy. Based on the tactical fire support requirements, positions must be chosen that ensure maximum fire support, efficient logistical operations, and protection from the elements. The terrain and weather's effects on communications further complicate the problem.

3. Movement. Movement is considerably more difficult. It is often the key to survival and success. Artillery commanders must be prepared to move their units across difficult terrain under extremely adverse weather conditions. Reconnaissance, selection, and development of positions are extremely important.

   a. Ground Movement. No commander should place his unit into a movement artery or position unless there are suitable areas for passing and stopping. Routes must be prepared with acceptable passing points cleared and turnarounds constructed. Military police (MP) support will be necessary to control MSRs during displacements. Batteries may have to carry dunnage to move effectively into and out of firing positions. Moving howitzers in deep snow is impossible until much of the snow has been removed from trails, tubes, and wheels. The positions must be prepared by snow removal equipment. Roads in cold weather operating areas have steep crowns and lack shoulders. Because shoulders are often snow-filled, it is not always possible to determine where shoulders end and ditches begin. Drivers will tend to move to the right shoulder when vehicles are approaching or attempting to pass. As vehicles slow and move to the right, they may slide laterally into ditches. This can result in weapons rolling over and prime movers getting stuck. Wreckers must accompany movement convoys. Chains are required for prime movers and artillery pieces. The M-198 tends to slide on snow or ice-covered roads. The RT-10,000 may be a consideration to assist in the movement of artillery pieces over extremely difficult terrain. However, the limited number of RT-10,000's will make them a precious commodity.

b. Air Movement. Displacement of artillery by helicopter should be avoided if possible. Signatures presented by helicopters in the LZ easily identify new positions. While moving batteries by helicopter avoids road and trafficability problems, it requires an increased reliance on helicopterborne logistical support in positions not accessible to roads. Batteries may become stranded because of competing helicopter lift demands or unsuitable weather conditions. If air movement is used, section equipment can be piggybacked. Section equipment and survival gear should be loaded on ahkios and carried inside the helicopter with the howitzer crew.

c. Self-Recovery. The ability to affect self-recovery without calling for outside support is imperative to ensure uninterrupted fire support and survivability. Drainage ditches are often parallel to foreign roads. Wheeled vehicles get stuck while attempting to negotiate these drainage ditches. The engineer materials and equipment that will be required to move towed howitzers over these ditches include:

- Dunnage, carried by each firing battery, which can be assemble to form bridging to span obstacles 15X15 feet.
- Additional wire cable or tow chains (assists in extracting vehicles from snowdrifts/ditches).

Generally, the winches on prime movers and forklifts associated with artillery batteries are too light to extract weapons or vehicles from ditches or snowdrifts.

d. Host Nation Support. Snow removal equipment can be procured several different ways. The host nation may assist U.S. Forces by providing direct support or the host nation may lease the equipment to U.S. Forces. Plan to integrate host nation support engineers into the snow removal plan if the host nation is offering support. Depending on the situation, the use of captured equipment may be the best option.

4. Survivability. The range of the M-198 (30 kilometers [rocket assisted projectiles]) may reduce the number of moves necessary to range targets. Severe weather, or difficult terrain/trafficability may decrease the number of moves that a commander may be capable of making. If frequent moves are impossible, other means of survivability such as hardening and camouflage must be stressed. Earth defenses of the conventional type are difficult to construct. Fortifying positions using logs or powder canisters necessitates the use of other tools (chain saws, engineer stakes, etc.) and leaves a greater logistical footprint. Making fortifications from ice or ice-crete (boxes filled with gravel and water then allowed to freeze) may be more readily used. Capabilities of host nation support engineer units must be integrated with the batteries to ensure development of hardened positions. Commanders must ensure that personnel practice sound communications security and electronic security. These techniques take on added importance when movement is restricted.

C. Weapons and Munitions.

1. Munitions Effects and Accuracy. Cold weather has many adverse effects on firing, responsiveness, effectiveness, and accuracy of the artillery. The basic procedures for firing remain the same, but cold weather affects the performance of the ammunition, fuses, and mechanical equipment. These effects must be understood to counter them to ensure effective artillery support. The weight of the 155 mm projectile is sufficient to crater most ice and heavy snow areas. Munitions effects and accuracy can be reduced by extreme weather conditions and deep snow (refer to Figure 3002-1). Rapid temperature changes may cover or expose family of scatterable mines (FASCAM), rendering them ineffective. FASCAM may come to rest in the snow at angles causing a nonkilling orientation of some of the mines, making them less effective.

2. Movement of Artillery Ammunition. Moving ammunition from the ammunition storage area to the firing batteries will require that convoys receive priority of movement on the limited MSRs. Coordination by the MPs will be required. Movement of ammunition requires:
   - Careful management of the MSR (a coordinated effort of both USMC MP and host nation liaisons/MPs).
   - Establishment of storage areas for the ammunition (requires snow clearing, camouflage, and hardening of positions).
   - Coordination between the artillery battalion and the combat service support element (CSSE).

Figure 3002-1: Cold Weather Effects on Munitions

CONDITION	EFFECT
Mountainous Terrain	• Mountain weather changes rapidly. Meteorological data downrange can be considerably different from where data was

	taken. This can affect the delivery accuracy. • Affect the ability of FO's to accurately locate targets.
Limited Visibility	• Reduces the effectiveness of white phosphorus as a marking round for CAS. Alternatives include illumination on the ground and 81mm mortar red phosphorus. • Reduces accuracy for laser-guided munitions, due to fog, ice crystals, precipitation.
Deep Snow	• Snow and ice will reflect lasers and can cause inaccurate delivery of munitions. • Absorbs the effect of point detonating fuses. • Variable time fuses have the potential to malfunction as air burst over snow. Favor mechanical time fuses. • Loose and powdery snow may preclude detonation of improved conventional munitions (ICM) grenades. • FASCAMS may settle into snow preventing area denial artillery munitions' tripwires from properly deploying or reducing the effectiveness of a remote anti-armor mine system.

3. Maintenance. During periods of extreme cold, malfunctions and broken parts will be more common. Weapons systems will need to be prepared for cold weather operations before deployment, following guidance in appropriate technical manuals (TMs). Maintenance in the cold will take two to five times longer. Contact teams will require protection from the cold. Tents may need to be erected over equipment/weapons systems and heaters used while being repaired. If heat is applied to a piece of equipment/weapons system, or if the repair is accomplished within a heated tent or building, sweating and differences in expansion/contraction will occur. Repairs cannot start until sweating has stopped.

4. Storage. Ammunition should not be placed inside tents or shelters that are so warm they will cause condensation. Projectiles, powder canisters, fuses, or primers may freeze when exposed to the colder outside air. This may damage propellants, increase difficulty in handling heavy projectiles, or prevent the proper mating of fuses and projectiles. Old firing positions can be effectively used as ammunition storage points.

5. Propellant. Cold weather will affect propellant characteristics. Proper handling and the continual updating of propellant temperatures will ensure effective gunnery.

6. Projectile Plugs. Projectile plugs should be left in place or fuses immediately mated with projectiles to prevent condensation, ice, or snow from accumulating in the fuzewell.

7. Powder Canisters. Powder canisters should remain sealed until just before use to prevent condensation.

D. Target Engagement. The maneuver commander established target engagement criteria and priorities depending on the enemy situation and the mission of the maneuver forces. Regardless of the climate, the engagement of individual targets with the most effective munitions and weapons system is a function of the forward observer, firing battery fire direction center (FDC), artillery battalion FDC, and the appropriate fire support coordination center based on the commander's guidance. Each of these agencies must be knowledgeable of the effects of the cold on range capabilities and munitions effects. A medium howitzer like the M-198 is a benefit with regard to target engagement because of its range. This provides the maneuver commander with a wide range of latitude for movement.

E. Reconnaissance and Selection of Positions. Position reconnaissance becomes more difficult and takes on added importance. Severe terrain, deep snow, and muskeg provide an unstable platform making artillery employment difficult. Off-road trafficability can be impossible for towed artillery. When locating suitable firing positions, Marines must consider dense trees, stumps, and deep snow. Off-road movement requires preparation by a bulldozer or other snow removal equipment that will either go along with or follow closely behind the advance party. In a fast-moving operation, artillery will stay on and shoot from dirt roads unless positions can be found to drive into and out of without preparation. Regardless of the tactical situation, weather conditions may keep artillery roadbound. Therefore, the route must be suitable to support firing along the way.

1. Planning. The artillery commander must begin his reconnaissance with a thorough understanding of the tactical situation and the GCE commander's intent. Based upon these considerations, he must use every available source to plan his reconnaissance. The numerous assets he can use to plan his ground reconnaissance include:

Aerial photography.
Map study.
Liaison with local personnel. This liaison often uncovers acceptable areas, which are not easily identified under a cover of snow.

A good basic rule is to NEVER POSITION AN ARTILLERY UNIT IN WINTER WHERE ONE CANNOT BE POSITIONED IN SUMMER! This should avoid all the problems which occur due to deep ditches, muskeg, lightly frozen ground, and tree stumps.

2. Potential Positions. Abandoned roads, industrial areas, quarries, and farmyards may be used as positions. The commander should look for places where indigenous personnel park heavy equipment and farm machinery. Supply/resupply considerations greatly influence selecting position areas. Always considered in position preparation are locations, state of communications facilities, and the proximity to the supported maneuver elements.

F. Continuous Reconnaissance and Preparation.

Reconnaissance is a continuous process concurrent with the preparation of positions. The battery commander during his reconnaissance and selection of position establishes the track plan for an artillery position. Ideally, position preparation is accomplished before advanced parties arrive. When possible, the higher level artillery command prepares positions one or two moves ahead of the firing units. Such a system greatly increases alternatives and provides for increased tactical deception with dummy positions and communications decoys. This continuous process is extremely strenuous for equipment operators and hard on snow removal equipment, but provides maximum flexibility and tactical support.

During the winter, it may be impossible to dig in positions, but parapets of snow and ice can be erected. During the clearing process, these parapets must be constructed to allow proper clearances, provide cover and concealment, but not impede the firing of weapons. In extreme weather, some type of heated shelter must be provided for personnel to periodically use to avoid cold injuries and attain adequate rest.

G. Engineering Support. An engineer detachment should be attached to the artillery battalion. This detachment should be tasked with snow clearing, hardening positions, and contributing to the local security. Engineers should be prepared to constantly improve and harden positions. Preparing positions can be a host nation support engineer function.

H. Battery Positions. The two primary methods of snow removal for firing position preparation are racetrack and driveway. Preparation of these positions is beyond organic capabilities of the artillery battalion. Responsibility can be assigned to support engineer units or host nations support engineers. Liaison with these units is a constant, ongoing process that must be rehearsed. Key in snow removal is to remove as little snow as possible and knowledge of the capability of the weapon system, the prime mover, and the state of training of the unit.

1. Racetrack. The racetrack is the least desirable method because of the amount of space and time needed for preparation. The racetrack is a large circular path with radiating driveways for the howitzers and support equipment (refer to Figure 3002-2). This plan is the standard method for preparing positions in temperate zones. In snow-covered terrain, it has many disadvantages.

It is difficult to camouflage because of the wide tracks that often must be made across pristine snow surfaces.
It is difficult to develop because of the several acres of fairly level ground required.

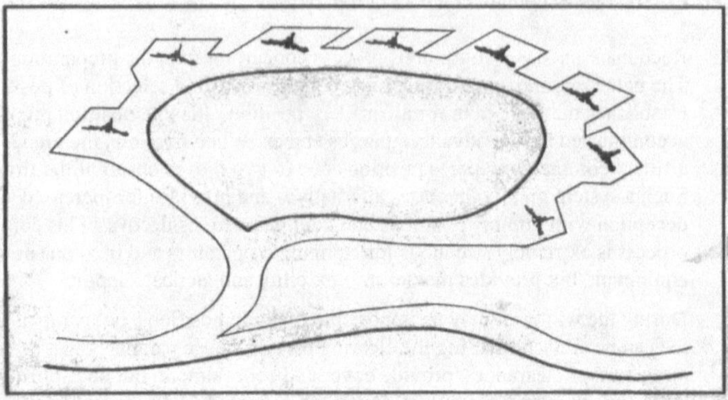

FIGURE 3002-2: RACETRACK METHOD FIRING POSTION

2. Driveway. The driveway snow-clearing plan is fast and easily developed (Refer to Figure 3002-3). A level stretch of plowed roadway replaces the racetrack. Many short driveways are dug/plowed into the snow on the shoulders of the road to allow individual pieces of equipment to get off the roadway. The driveways for howitzers are dug so that they point roughly at the enemy. Artillery batteries will need to carry dunnage to fill the ditches formed at the edge of the road. The driveway has the following advantages:

Time required to dig a battery position using driveways is less than one-third.

Camouflage is easier as white nets readily blend with the irregular, discolored snow surfaces caused by repeated highway plowing.

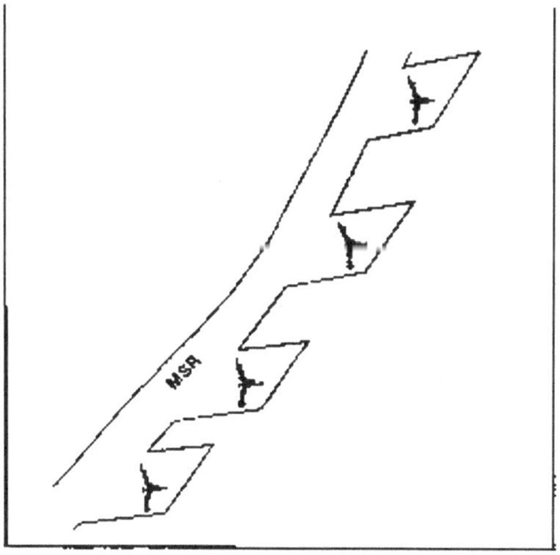

FIGURE 3002-3: DRIVEWAY METHOD OF POSITIONING

I. Supplementary Positions. Old positions should be retained for use as supplementary positions. They lie within the artillery fan of supported infantry and will have had their surveys completed. They can still be used to fire missions should primary positions need to be abandoned because of counter-battery fires. GCE/CSSE will want to use these old positions as ammunition, logistical, and supply points. Commanders are cautioned that these positions must not be occupied while they can still be used as supplementary firing positions.

Split Battery Positions. Battery positions will often need to be nontraditional, split into multiple gun positions, using whatever open space is available which will support the guns in accomplishing mission.

J. Surveying. Adverse weather can greatly reduce the capabilities of survey teams. Locating bench marks/survey control points in snow-covered terrain is difficult. Like other Marines in the cold, survey teams will need over-the-snow movement skills to conduct their mission. Often, known points with established locations and elevations may be used; e.g., bridges. Staking survey markers can be difficult in frozen terrain or deep snow. Once in place, these markers can easily be obscured by falling or drifting snow before firing batteries arrive. Survey markers will occasionally fall over as the metal stakes absorb heat from the rays of the sun and the supporting snow melts. The position azimuth determining system (PADS) mounted on the HMMWV generally works well in the cold. However, success with PADS depends on trafficability.

3003. Tanks and AAV Operations

The combat missions assigned to armor units in cold weather operations are the same as those in a temperate environment. Deep snow and extreme cold weather present employment problems that require aggressive leadership, specialized training and snow clearing assets. Extensive training prior to operating in a cold weather environment is encouraged. Tracked vehicles need to be prepared to drive in both snow and icy conditions. Mud will become troublesome during the change in seasons.

A. Effects on Personnel. Personnel need to be aware of environmental effects on the body. In particular, special considerations need to be in place to combat snow blindness and frostbite. UV sunglasses are necessary to combat snow blindness. Contact gloves need to be worn at all times during extreme cold conditions around vehicles. Cold metals and fuels put bare skin at higher risk for cold weather injuries. SOPs and discipline will be required to prevent unnecessary injuries. Heaters will help keep personnel warm when inside the vehicles. Marines should always be prepared with the proper clothing and equipment to survive if required to abandon their vehicle.

B. General Mobility Considerations. Normal mechanized operations can occur in snow depths up to 3 feet of snow. Commanders must be aware that snow and ice will get packed into the tracks. This could possibly result in a track being thrown. Furthermore, once stopped, snow and ice needs to be removed from the tracks, suspension idler wheels and sprockets to prevent the formation of ice. If operating in arctic regions, tundra may freeze over and allow vehicles to cross. Travelling in columns over tundra is not recommended. The repeated weight and vibrations could break through the frozen crust, resulting in the loss of a vehicle. Crossing ice is possible if the ice is thick enough to support the weight of the vehicle. Refer to Section 3004: Combat Engineer Operations for information regarding ice support capabilities.

1. Tanks. Tanks will operate effectively in up to 3 feet of wet snow and up to 4 feet of dry powdery snow. Tanks require 11.3 pounds per square inch (lb/in^2) to achieve traction. If this is reduced (as in deep snow or on ice), tanks will lose mobility. Removing the rubber pads on selected tracks can enhance mobility.

2. AAVs. The AAV can negotiate up to 35 inches of snow during cross-country movement. Normal speeds can be maintained if a packed trail has been formed and the surface has compacted into a hard mass resembling well-packed wet sand. Commanders should understand that elevation will decrease the overall vehicle capabilities. For example, speeds and traverse angles (60% forward grade and 40% side grade) will be reduced. Refer to FMFM 9-2: Amphibious Operations to learn more about AAV capabilities. Depending on road conditions and local restrictions, AAVs should employ the X-cleat or operate without track pads for improved traction on ice or hard packed snow. Halts should be made during movements to clear snow and ice from tracks, suspension idler wheels and sprockets. Caution must be exercised when conducting AAV operations in icy waters. Off shore ice formation may limit access to the beach. Efforts must be made to prevent Marines form being exposed to water and spray.

3. LAVs. LAVs are capable of moving through snow up to a depth of 3 feet. Deeper snow, iced snow banks and narrow roads in coastal cold weather areas will restrict the LAV's off road mobility.

C. Mechanized Vehicle Operations. Although the commander must understand the difficulties associated with mobility, the tactical considerations for mechanized forces will not differ from a temperate environment.

D. Recovery Operations. Recovery operations should begin as soon as possible after a vehicle becomes disabled to prevent freezing in. For this reason, recovery equipment must be kept as far forward as possible. The addition of recovery vehicles to a convoy may become necessary in the cold weather environment.

E. Maintenance. Heated facilities are necessary to repair vehicles. At temperatures below -40° F, maintenance will require up to 5 times the normal amount of time. Special emphasis must be placed on the timely performance of required organizational maintenance. Increased stocks of repair parts are necessary. The greatest increase is for electrical components and those parts that depend on lubrication for long life.

Winterization kits must be installed according to the manufacturer's specifications found in TMs. These kits include heaters and wind deflectors. Chains are installed on four of the six drive wheels of the LAV according to TMs. They cannot be applied to wheels that are adjacent to each other, as the chains will interfere with one another.

Remember to have fluids and lubricants changed according to lubrication orders and TMs prior to being placed in a cold weather environment. In theater the appropriate fuel mixes are the responsibility of bulk fuel units. As temperatures change to warm weather some degradation can be expected from winterized vehicles.

F. Fuel Consumption. Anticipate higher than normal fuel consumption rates. Longer idle times are necessary to keep equipment at operating temperatures. Fuel filters must be drained more frequently than usual to prevent freezing. Keep fuel tanks as full as possible and purge the fuel cells to remove condensation with the transfer pump. The water / fuel separator must be tested and adjusted to purge for 6 to 10 seconds.

G. Effects of Extreme Cold. Vehicles become cold-soaked at -30° F. If this occurs with tanks (M-1A1), the purge pump pre-heater is used for cold starting procedures. Engines should be operated and tanks exercised periodically to maintain their operational capability. Starting and warm-up time is also increased, and may approach 2 hours in temperatures of –50° F. Special consideration must be taken to ensure battery life is maintained. When vehicle temperatures fall below –50° F, warming tents for water and oil containers are required to keep them usable.

3004. Combat Engineer Operations

A. Preparation. Combat Engineer units tasked to execute operations or exercises in a cold weather environment should prepare by:

- Conducting proper cold weather training in order to be prepared accomplish any engineer mission.
- Modifying organizational equipment to support cold weather operations.
- Acquiring and issuing special-purpose equipment as follows:
  - Ice-measuring rod
  - Ice auger/axe bar
  - Chisel/spud
  - Ice saw
  - Weighted depth cord
  - Probes
  - Belay rope
  - Axe
- Emphasize aggressive leadership and develop a proper mindset that is needed to succeed in this environment.

B. Reconnaissance. In a cold weather environment reconnaissance patrols will often require the assistance of combat engineers to verify accessibility and capacity of roads, LZ's, beachheads, and ice. This will aid the commander in making decisions in regards to trafficability of planned routes and supportability of the operational plan.

1. Route Selection. Waterways such as rivers and lakes can be obstacles during the spring and summer months but can become trafficable and an asset during the winter months. Full use of all intelligence available through map, ground, and aerial reconnaissance is necessary for proper route selection. Route selection criteria will vary by season. Regardless of the season, the need for roads will not be eliminated by over the snow vehicles.

2. Engineer Ground Reconnaissance. Engineer reconnaissance includes observation of soil, snow cover, vegetation, ground water, surface water, ice thickness, road surface conditions, sources of local construction materials as well as the conditions of alternate routes. The purpose this reconnaissance is to:

    - Verify all information previously collected

- Check all possible routes for natural and man-made obstacles (for example: avalanche debris, mines, ice obstacles, and others)
- Select the best site or route to accomplish mission

C. Snow and Ice. Collect as much data and information as possible in the AO in regards to snowfall and ice growth. Both are equally important and can be used as an advantage or be a disadvantage to friendly forces.

1. Snow. As snow accumulates it will reduce the mobility of both man and machine. Heavy accumulation of snow can result in avalanches. Engineers need to be aware of the slope angles and aspects that favor avalanches in their AO. This enables the commander to use the forces of nature to his advantage. Refer to Appendix A: Avalanche Danger, Recognition and Rescue to better understand the destructive capability of snow.

2. Ice. Fresh water freezes at 32° F and salt water freezes at 28° F. Ice is relatively strong and has a varying degree of toughness and high bearing power. As temperatures drop, the strength increases rapidly from the freezing point to about 0° F. From this temperature the strength of ice remains fairly constant despite lowering temperatures. Refer to Figure 3004-1.

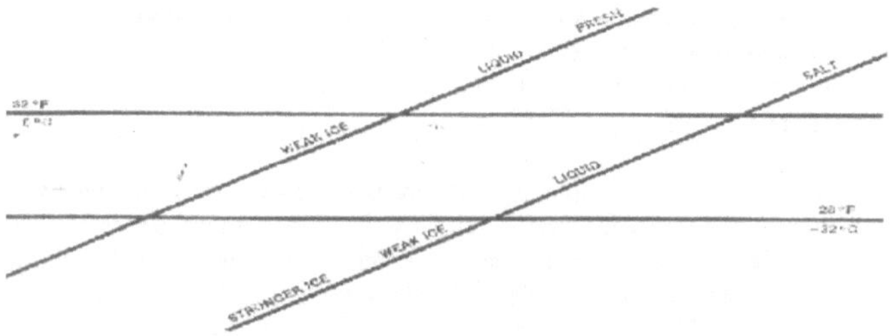

FIGURE: 3004-1

D. Strength of Ice. Generally, freshwater ice is more uniform and stronger than sea ice. The strength of ice is dependant upon the following conditions:

- Structure
- Water Purity
- Process with which Ice was Formed
- Has the Ice Undergone Freezing and Thawing Cycles
- Orientation of Crystals
- Temperature
- Thickness
- Snow Cover Over Ice
- Water Currents
- Support Beneath.

- Age

1. Indicators of Weak Ice. There are two classifications for weak ice: rotten and unsupported.

   a. Rotten Ice is caused by a thaw or by incomplete freezing. During winter months bogs, rotting vegetation, or even sewage typically causes this. Regardless rotten ice occurs when a contaminant of some form disrupts the freezing process. Generally rotten ice is dull and chalky in color and is brittle. Rotten ice has limited to no strength and should not be used.

   b. Unsupported ice is the second classification of weak ice. This form of ice occurs when there is space between the ice and water. It is normally found in areas where the water table has fallen due to tidal action. If operating in an area where dams are prevalent, understand that this water can be drained causing a severe case of unsupported ice.

2. Indicators of Strong Ice. Generally, a thickness of 14 to 20 inches of waterborne freshwater ice is necessary for safe passage of heavy equipment. A safe rule of thumb for ice thickness required for armored vehicles is 16 inches of waterborne ice can support 16 tons, and each additional inch can support an additional ton. Blue ice is by far the best and strongest. Normally the color is light blue or green in shallow areas and black over deep water. Some cracks are visible but these do not weaken the ice. They run in the same direction as the current. Refer to Figure 3004-2 for the load capacity of good quality blue ice for military equipment and personnel.

3. Testing Ice. One of the most important responsibilities of the engineer in this environment is to test ice for trafficability. Prior to testing the ice and getting readings, the engineers conducting the reconnaissance must check for mines, ice obstacles, tank traps, enemy demolitions under the ice, and evidence of NBC use by the enemy. The following readings must be taken to paint the clearest picture possible of the waterway:

   - Measure ice thickness
   - Measure depth of snow on ice surface
   - Determine how ice is attached to the banks
   - Determine the slope and approaches of the banks
   - Measure the width and depth of the water
   - Measure the fastest and slowest current speeds
   - Determine capacity of ice in accordance Figure 3004-2
   - Gather data on the theoretical growth of the ice for the region.
   - Determine requirements to make the waterway passable
   - Determine availability of reinforcing materials, if required

OBJECT	Weight in Tons	Minimum Ice Thickness in Inches	Minimum Distance between Objects in Feet
Man on Foot	.1	1.3	10
Man on Skis	.1	1.3	10
Horse	.3	1.4	11.6
Motorcycle KLR 250	.12	1.4	11.6
Horse Drawn Vehicle	2.5	6.3	53
Horse Drawn Sled	2.5	6.3	53
Snowmobile	.5	2.8	23
BV 206 (combat loaded)	7	10.5	88
HMMWV (M-7178)	3.5	7.4	67
APC	2.5	6.3	53
3 Ton truck w/ Load	7	10.5	88
M9 Armored Earth Mover	27	20.8	173
5 Ton Truck M-813A1	10.2	12.8	107
M3-450 Bulldozer	5.5	9.3	78
LAV	14.2	15.1	126 feet
AAVP7A1	27 (combat loaded)	21	175 feet
M1-A1	63	36	300 feet

FIGURE 3004-2: ICE SAFETY TABLE

To determine the minimum solid ice thickness required to support use the formula:

H= 4 x SQRT (P)

H= equals ice thickness in inches
SQRT= Square Root
P= The load or gross weight in tons

To determine distance in Feet:

100 times the ice thickness then divide by 12 to determine feet
100 x H / 12= Distance in Feet

E. Mines.

1. Mine Laying. Snow, ice, frozen ground, and low temperatures affect mine-laying operations. Burying mines in a frost layer of more than 8 to 10 centimeters may be prohibitive. Mines must be placed on top of the ground. Snow or ice can prevent sufficient pressure from being put on the mine to cause detonation. Water seeping into and around the pressure-firing device may freeze and prevent detonation. Plastic laid over the top of the mine will prevent this type of failure. Mines should be painted white if possible for camouflage purposes. Some special considerations that must be adhered to during winter conditions are as follows:

a. Prepare the M605 fuse for the M16A2 anti-personnel mine by cleaning off all the packing grease with diesel. The diesel will effectively clean and lubricate the fuse as well as making it water-resistant. Replace both safeties with safety pins to allow for easier arming and disarming.

b. Prepare the M603 fuse for the M15 anti-tank mine by dipping the fuse in a lacquer to make it water-resistant.

c. Use of anti-handling devices is highly discouraged in this environment due to the high probability of premature detonation due to the ever-changing conditions. (Melt-freeze and shifting of snow pack)

d. Employing a sub-surface mine in this environment leaves a distinguishable signature. Try to use the forces of nature to cover your tracks. For example, lay your mines during or just prior to snowfall.

e. Employing anti-personnel (AP) mines and anti-tank (AT) mines in deep snow requires a firm bearing surface such as frozen ground, boards, or rock bases. When employing AP mines, select likely ski trails. Because of the low surface pressure while on skis, only place the mine about one inch under the snow or attach tripwires (painted white for camouflage purposes). Place mines on a downhill slope just around the bend of a trail because enemy ski patrols are likely to be well-spread out, it is often effective to link a series of explosives with the mines at intervals of five meters down the track to allow for better effects.

2. Arming. Arming mines in quantity is difficult in low temperatures. It is recommended to prepare and handle all ordnance in a warm shelter if possible where Marines can be more efficient.

3. Family of Scatterable Mines. Air and artillery delivered mines are quick and easy to employ. However, they may not settle upright in the snow and may malfunction making the minefield ineffective and leaving a possible obstacle for friendly forces operating in the area.

4. Countermine Operations. Countermine tasks in this environment are also affected by the cold. Mines placed in snow cover can be plowed away to clear a route. Detonation and breaching can be more difficult due to freezing temperatures, frozen ground, and concealment of the snow cover. (Refer to FM 5-102, *Counter-mobility.*)

F. Obstacles. Commanders should allot more time for Marines to construct obstacles in cold weather. When developing obstacle plans, one must consider what effect changes in weather will have on the plan. For example, if temperatures rise significantly, many areas that were solid ground may become untrafficable, such as rivers and like. The converse is likewise true. If temperatures fall causing rivers and lakes to freeze these may become new avenues of approach for the enemy. These areas can and should be covered by demolitions or artillery.

The engineer should never fail to use the natural obstacles that the environment offers. Icy slopes and fallen trees can disrupt and channel troop movements. Leeward slopes with heavy deposits of snow can be rigged with explosives in order to catch enemy troops in the avalanche runout zone. Barbed wire and concertina are still effective on snow.

G. Field Fortifications. Successful construction of hasty/temporary field fortifications can be accomplished with military explosives. Fighting positions can be constructed of ice and snow. However, in the event of a thaw, these positions lose strength, become wet, and are difficult to fight out of. Snow and ice will break down under sustained fire. Reference Appendix B: Fieldworks and Camouflage for more information.

1. Survivability. Constructing field fortifications and fighting positions in frozen ground is difficult. More time is required to dig in regardless of the tool or method used. Unless removed from the area or sandbagged, the spoil leaves a prominent signature on the snow. Marines must be able to construct field fortifications on snow and frozen ground with available materials. Units should conduct training in accordance with FM 5-34: Engineer Field Data that focuses on employing hasty field fortifications. Effective expedient techniques can be used to build above ground positions using snow. Some factors to consider when building above ground positions of snow are:

Compacted snow will stop or slow projectiles and fragments
Compacted snow is easier to move than the lightest unfrozen soil
Snow on overhead will reduce heat signature
AT weapons with shaped charges may fail if the target is snow-covered.

FIGURE 3004-3: SNOW WALL THICKNESSES FOR STANDARD MUNITIONS

Snow Density (1 lb / ft^3)	Projectiles	Muzzle Velocity	Penetration (feet)	Required Minimum Thickness (feet)
18.0 – 25.0	Grenade		2.0	3.0
11.2 – 13.0	5.56 mm	3250	3.8	4.4
17.4 – 23.7	5.56 mm	3250	2.3	2.6
11.2 – 13.1	7.62 mm	2750	13.0	15.0
17.4 – 23.7	7.62 mm	2750	5.2	6.0
25.5 – 28.7	7.62 mm	2750	5.0	5.8
19.9 – 24.9	12.7 mm	2910	6.4	7.4
	14.5 mm		6.0	8.0
28.1 – 31.2	70mmHE AT	900	14.0	17.5
31.2 – 34.9	70mmHE AT	900	8.7 – 10.0	13.0

NOTES:

- These materials degrade under sustained fire. Penetrations given for 12.7 mm or smaller are for sustained fire (30 continuous firings into a 1X1 foot area.
- Figures given for 70 mm HEAT are for RPG-7 fired into machine packed snow
- HE grenades produce small, high velocity fragments which stop in about 2 feet of packed snow. Effective protection from direct fire is independent of deliver method, including weapons like the AGS-17 or Mk-19 40mm. Only armor penetrating rounds are effective.

a. Excavation is difficult in frozen ground. Explosives are effective at breaking through the frozen crust. Large quantities would dig in an entire unit. Charge calculations cannot be made directly from data in FM 5-25: Explosives and Demolitions. Because of the variations in moisture content, soil type, vegetation, and changes in explosive charge properties caused by cold temperatures; demolition computations should be made based on experience. Test shots will be necessary in most cases.

b. Hasty firing positions and trenches are built in the snow reinforced with available material such as ice, wood, or branches. A minimum of 135 centimeters (4 ½ feet) of solid packed snow is required for adequate protection from small arms fire.

c. Gaps in the defense should be reinforced. Artificial barriers like wire entanglements, abatis, minefields and road craters are possibilities.

2. Improvised Defensive Positions. Consider the following when constructing defensive positions:

a. If the snow is deep. Communication trenches should be dug at a minimum to better camouflage the position. If time and conditions permit, constructing overhead cover with snow above the communication trenches will reduce heat signature of personnel operating in the trenches.

b. Forested areas. Measures must be taken to protect defensive positions against forest fires that may be set deliberately.

c. Dummy positions are especially effective in winter, and should be used extensively to deceive the enemy both from the ground and air.

d. Wire barriers, although practical, tend to lose their effectiveness as snow deepens unless they are lifted and reestablished. They will also require continuous surveillance to ensure the enemy does not mine or booby-trap them.

e. Sandbags filled with sand or snow are effective and provide speedy construction of defensive positions in frozen ground. Water poured on the bag freezes and improves on their protective qualities for the duration of the cold weather.

3. Construction Materials. Field fortifications can be constructed out of natural materials found in cold weather areas.

a. Snow. Dry snow is less suitable for the expedient construction than wet snow because it does not pack as well. Snow piled at road edges after clearing equipment has passed hardens within hours even at very low temperatures.

b. Ice. The initial projectile stopping capability of ice is better than snow or frozen soil. However, under sustained fire, ice rapidly cracks and collapses.

c. Ice Crete. Ice crete is a mixture of sand, soil, chipped ice or gravel, and water. Material made of gravel-sand-silt aggregate saturated and poured like cement is also suitable for constructing positions. Once frozen, this material has properties like concrete. Sub-freezing temperatures are required to construct and maintain ice crete.

d. Frozen/Unfrozen Soil. Frozen soil is three to five times stronger than ice and increases in strength with lower temperatures. Frozen soil has great resistance to impact and explosion. Resistance is an especially valuable feature for constructing positions. Unfrozen soil from beneath the frozen layer is sometimes used to construct a position. This must be accomplished quickly before the unearthed soil freezes. Also, once the permafrost layer is broken, the creation of underground positions will be easier.

H. Cold Weather Effects on Demolitions. Extra care must be taken in regards to the storage, handling and preparation of explosives. Refer to Appendix B: Fieldworks and Camouflage for more information.

CHAPTER 4
AIR OPERATIONS

4001. Preferred Methods Of Marking Targets For Close Air Support

Some target marking methods are equally useful in both cold and temperate climates. Using prominent terrain features subject to ensuing snowfall and cover, signal mirrors, or air panels for common reference points are viable methods as long as they are available. Crossed tracer fire from machine guns is effective, but as in temperate climates, this method puts the survivability of the machine guns in question. Preferred methods that have unique application or increased effectiveness in snow-covered terrain and the cold are:

A. Dark Smoke. Dark smoke as a marking method will be more viable in snow covered terrain. In the past, Marines have used burned houses or already burning vehicles as a common reference point. The darker the smoke, the better it contrasts with the snow. Burning tires or petroleum, oil, lubricant fires make excellent dark smoke. Fewer mechanized and motorized vehicles will be used in cold weather environments than in temperate environments. Obviously, fewer vehicles will be available to be destroyed and thereby to create smoke. There will still be a great deal of smoke on the battlefield, but not as much as in temperate climates.

B. Illumination. Illumination suspended from a parachute over the target provides a distinct mark both during daylight and darkness. The source of the illumination can be artillery, mortars, or the M-203. The height of the illumination must be coordinated so that it does not interfere with the attack heading of the aircraft.

C. Colored Smoke.

1. Currently only white smoke is available in 155mm rounds. Colored smoke is available in 105 mm rounds, which the Marine Corps no longer uses. However, you may see other nations use the 105mm rounds and hence the following information is included. The round is equipped with a mechanical time fuse and several base ejecting smoke canisters, the height where the round is activated can be adjusted to provide colored smoke streamers form a point over the target down to it. The round activation height should be as low as possible but still able to get the desired effect. If the round is activated too high, the dispersion of the canisters when they hit the snow will be to great to pinpoint the target. Increased smoke from the canisters can also obscure the target. Timing of the activation of the round over the target should be such that it occurs just before the aircraft reaches the peak of its pop where it will invert to acquire the target.

2. Colored smoke is also available from smoke grenades and hand-held popup canisters. These can be used to make friendly positions and establish common reference points. Smoke grenades must have a flotation device to prevent their sinking into the snow. Smoke must be activated early enough so that it creates a signature sufficient to be acquired by the pilot as he inverts in the popup.

D. Ice Fog. High explosive rounds with variable time or mechanical time fuses can be used as expedient marking method in extreme cold (-25F). The round can be timed to explode over the target to create a cloud of ice fog. The cloud may not be very distinct but can suffice in the absence of other more viable methods.

4002. Target Methods Degraded by Effects of Cold Weather

Many methods used in temperate climates are ineffective because the mark disappears in the snow. Systems used to designate the target or to provide a known reference point are reduced in effectiveness. Problems are normally associated with the initial mark. (Munitions from the lead aircraft usually created a distinct reference point for subsequent aircraft.) Some problems are:

A. White Smoke. The white smoke produced by hydrochloroethane (HC), white phosphorous (WP), and the M 825 improved smoke does not contrast with the snow and cannot be seen.

B. Smoke Canisters. Burning smoke canisters quickly melt into the snow. If the snow is deep, the smoke will be smothered.

C. Illumination. If illumination rounds are timed to fire just before hitting the deck, they will provide a good mark. However, illumination rounds melt quickly in the snow. If the snow is deep, the illumination cannot be seen.

D. Laser Systems. Laser systems will operate well in a cold weather environment if they are given adequate warm-up time. Moisture is a problem on all laser systems operating in cold weather operations. Equipment must have adequate moisture seals. A slight over pressurization of equipment housings using dry nitrogen and desiccant absorbers will help maintain a low dew point. The optical window must be kept dry and ice-free. In addition, the laser system should provide a flashlamp/element preheating as well as anti-icing capabilities. **Use extreme care when operating laser systems in cold weather environments. Ice and snow are sources of high reflectivity that could cause eye damage**. In addition, overspill and underspill of the laser will present more of a problem due to the reflectivity of the winter conditions.

4003. Planning for Helicopter Operations

The helicopter is the single best tactical mobility asset available to Marines during cold weather operations. It can move you farther and faster than any other means of transportation. The helicopter has limitations. The greatest of which are the lack of dependability due to unpredictable weather and the extreme difficulty of performing maintenance in the cold. Additional maintenance personnel and maintenance shelters may be required. This means that the unit leader must always have an alternate movement plan to get to the destination in time to accomplish the mission! Use the helicopter whenever possible, but beware of becoming totally dependent on it. Helicopters are vulnerable targets due to their size and speed. The following items should be considered in employing helicopters in a cold weather environment:

A. Reduction in Operational Tempo. Remember that everything takes longer in a cold environment. It takes the mechanics longer to fix, fuel, and do routine maintenance on the aircraft. The aircraft may have more maintenance problems due to the cold weather. Fuel lines, hydraulic systems, and electrical systems require more maintenance to operate in extreme cold temperatures. In addition, it will take longer to load, unload, stow equipment, hookup seat belts, and conduct approaches and departures into Landing Zones (LZ's) in a cold weather environment.

B. Vulnerability in the LZ. Delays in the LZ will make helicopters particularly vulnerable targets to both direct and indirect fires. Helicopters often create large snow signatures when conducting landings and takeoffs into a snow covered LZ, this makes friendly forces extremely visible to enemy observation.

C. Temperature and Altitude. As temperature and altitude increase, helicopter performance decreases. This affects not only payload capability of the helicopter but also time on station, airspeed, and maneuverability. Decreased temperatures will not offset the effect of increased altitude when operating in high mountainous terrain, therefore helicopters will not perform as well as they do at sea level. Not all cold weather operations will be conducted in mountainous terrain. Most USMC cold weather operating areas are maritime and although mountainous, are not the high alpine-type terrain.

D. Weight/Bulk Load. In a temperate climate, a combat-loaded Marine averages 225 lbs. In a cold weather environment, the average Marine weighs 300lbs because of the increased weight of cold weather clothing, equipment, and rations; a difference of 75 lbs.! It will take 1 ½ normal seating space for a Marine with a full cold weather gear, thus reducing actual troop space.

E. Weather. Mountainous or arctic terrain is compartmentalized and is characterized by rapid change. Weather may be good in the pickup LZ and bad in the destination LZ. Consequently, commanders must have alternate plans for insertion and extraction if possible because weather will be unreliable.

F. Rotor Wash Identification and Visibility. On landings and takeoffs, helicopters re-circulate large snow clouds; i.e., snowballs that can be observed form considerable distances. The rotor wash of helicopters flying nap-of-the-earth will remove the snow from the trees. This can provide the enemy with indications of flight routes and also help them locate friendly units. These techniques can also be used to create deceptions.

4004. Assembly Areas

Assembly areas should provide security, concealment, dispersion, and a windbreak for Marines. If boarding delays occur, warming tents may be necessary. Anytime Marines wait longer that 40 minutes; they should erect warming shelters. This period may be substantially shorter in extremely cold temperatures or under severe wind-chill conditions.

4005. Safety Considerations

Marines must understand safety considerations to reduce cold weather and mountainous environment helicopter operation hazards.

A. Frostbite. Frostbite is a constant danger due to the combination of wind-chill and cold temperatures. Use the buddy system to check for signs of frostbite.

B. Rotor Blade Hazards. In deep snow-covered LZs, helicopters may sink into the snow. This reduces the rotor-blades-to-surface clearance (Refer to Figure 4005-1). In sloping LZs, do not approach the helicopter form the upslope side as rotor-blade-to-surface clearance is further reduced. High winds while debarking can blow Marines and equipment back into the helicopter tail rotor blades. Tail rotor blades will also be much closer to the surface. Using the ahkio huddle-loading method will eliminate this danger.

Figure 4005-1: Increased Rotor Hazard / Potential Obstacles Below Snow Surface

C. Cargo Ramp Problems. In deep snow, the ramp of the helicopter may not lower enough to debark/embark Marines. The helicopter may have to lift off and move forward so the ramp can be lowered into the first landing depression in the snow. Marines must be aware of reduced head clearance and constant slipping hazards from ice in the cargo ramp area. The hydraulically operated ramp can easily catch and crush Marines.

D. Ice Shedding. If a helicopter has just flown through or is experiencing icing conditions in the LZ, there may be significant ice shedding hazard. Ice that accumulates on rotor blades will shed in many pieces and become flying projectiles. During ice shedding, Marines should stay in a staging area. If ice shedding occurs during offload/onload, it is safer to stay down low and inside the rotor arc, as in the ahkio huddle, than to try to move away form it. If icing conditions persist, it may be necessary to shut down the helicopter.

E. Unprepared LZs. When landing in an unprepared LZ, the fuselage will float on the snow's surface. Landing points should be probed and tramped down if possible to determine possible obstacles. Refer to Figure 4005-1. When terrain beneath the snow is uneven, the landing gear may or may not come to rest on the ground. Uneven landing could lead to dynamic rollover. The pilot may carry power to prevent settling into the snow. The resulting rotorwash may interfere with the onload/offload.

F. Dynamic Rollover Damage. If a helicopter settles in or breaks through a snow or ice surface, the helicopter may be a danger of dynamic rollover. Dynamic rollover is a dangerous condition where the helicopter could rollover onto itself due do a landing gear/skid coming in contact with the earth while power is being applied to the aircraft. This gear becomes a fulcrum point and the rolling motion is too fast to counter act by the pilot, which causes the helicopter to roll over. Therefore, it is extremely important to pack the LZ if time is not critical.

4006. LZ Brief

A pre-landing briefing between the ground unit and the helicopter unit is necessary if loading/unloading is to be conducted is a quick, efficient, and safe manner. The following information should be included:

- Description of LZ
- Wind direction and estimated strength in knots.
- Snow pack or Ice pack depth and whether the snow is packed or not.
- Obstacles or suspected subsurface obstacles
- Any special consideration that will delay embarkation significantly.
- Last know enemy position
- Loading method
- Method of marking the LZ

4007. LZ Selection

LZ size is determined by the number and type of helicopters to be employed. Landing a helicopter in a small or restricted LZ requires employing a precision type approach that may expose the helicopter to enemy observation and fire. The final approach to landing will be governed by the ability, experience, and judgment of the pilot in command. In blowing snow, the size of the LZ will have to be increased. Generally, the GCE unit leader will select the LZ and determine where the ahkio huddles or external loads will be staged. He must be thoroughly familiar with the following landing zone requirements.

A. Size. The minimum size of a snow-covered LZ for a CH-53, CH-46, or MV-22 is 150 by 150 meters. For an UH-1N or AH-1W, the minimum size is 100 by 100 meters. When a LZ has multiple landing points, all landing points should be a minimum of 150 by 150 meters apart. Refer to Figure 4007-2.

B. Approaches and Exits. The perimeter of the LZ should be clear of obstacles over 25 feet tall. Otherwise the LZ should be doubled.

C. Wind Direction. The wind determines approach and departure direction. Helicopters normally take off and land into the wind. To avoid the interference of the blowing snow cloud, helicopters will off center about 15 degrees (refer to figure 4007-1) to allow the snow cloud to clear.

Figure 4007-1: Downwind Approach

D. Ground Surface. The ground surface should be as level as possible. Small rocks, tree branches, etc should be cleared if possible so as not to create flying debris in the LZ.

E. Ground Slope. Terrain that slopes more than 8 degrees is usually considered too steep for helicopter landings due to dynamic rollover characteristics of all helicopters.

F. Concealment. LZs should be selected that conceal both the helicopter and the snowball signature from direct or indirect enemy observation. The white snowball that develops from the rotor wash can be observed up to 30 kilometers away.

G. Obstacles. The unit to be loaded should look for obstacles that may be hidden under snow. Obstacles that are hidden are potentially dangerous to the helicopter. Probing the LZ should be conducted to find tree stumps, large rocks, etc. which could rupture the skin or fuel tanks on the bottom of the helicopter (refer to Figure 4005-1). Additionally, minimum safe distances form obstacles/other landing points must be considered (refer to Figure 4007-2).

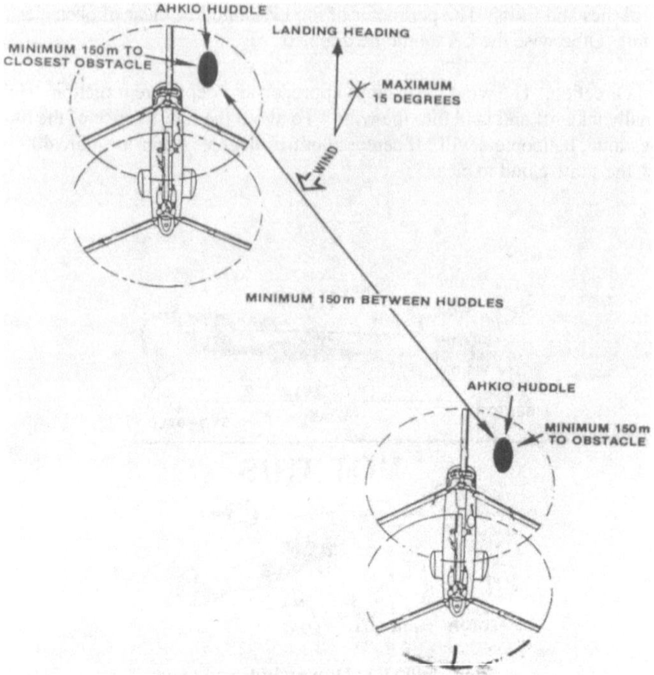

FIGURE 4007-2: Minimum Safe Distance Between Landing Points (Other Helicopters) or Between Helicopters and Obstacles

H. Snow. Depth and consistency of snow will have a major impact on LZ operations. Loose snow will blow and make it hard for the pilots to land safely. Hard or crusted snow may break up and become a hazard to Marines.

I. Lakes and Rivers as LZs. Commanders should include lakes and rivers as alternate LZs. Frozen lakes and rivers make excellent LZs since they are level and have little loose snow due to the scouring winds. Helicopters may skip on the ice during takeoff and landings; however, wind gusts may also blow helicopter while on the ice. CH-53, CH-46, and MV-22 need 15 inches of ice to conduct operations. UH-1N and AH-1W need 8 inches of ice thickness.

4008. Designation of Landing/Loading Points

Generally, a representative of the unit to be loaded will designate the landing points. Timing of approaches and wind direction can be used to compensate for blowing snow during multiple helicopter operations. Close coordination at the LZ brief will be necessary between the squadron and the unit to be lifted. Landing points should be selected so that the helicopter lands into a slight crosswind. This will make the snowball blow away for the intended points of landing for the following helicopters.

4009. LZ Preparation

Marines should make every effort to walk through the LZ to determine snow depth and appropriate locations for helicopter landing points.

A. Packing the LZ. Packing the LZ makes it easier for a pilot to find the landing point and for the Marines to move about. This consideration is particularly important when conducting external operations. Packing takes more time and the possibility of detection by the enemy may be increased.

1. Time, conditions, and tactical situation permitting, pack an area at least 50-meters square for each landing point. The area should be packed uniformly so that one wheel will not sink and cause the helicopter to land unevenly.

2. Over the snow vehicles are the most effective for packing LZ quickly. Marines on snowshoes, skis, or boot packing method can also be used but is more time intensive and exhausting.

B. Marking the LZ. Marking the LZ and the landing points is critical. The white snow-covered zones provide a difficult background for the pilots. The local white out hazards created by blowing snow when helicopters land will obscure the rotor arc if the helicopter slows to near zero airspeed in a hover. Reference points must be visible at al times. No-hover landings should be made so that the pilot retains a reference point with the ground. (No-hover landings keep the snowball behind the helicopter.)

1. The LZ can be marked using conventional panels and lights, by using rescue-survival dyes, dirt sprinkled in the snow, small green tree boughs, or any dark material.

2. A smoke grenade is an excellent way in which to show the pilot wind direction at the LZ. Place the smoke grenade on an object to prevent it from sinking into the snow. Do no use white smoke, as it will blend in with a winter environment.

3. Use an ahkio huddle to mark the landing points. The huddle should contrast in color to the background in the LZ. Individuals should remove overwhites, wear a protective face mask, and be sure no bare skin is exposed to the rotor wash.

4010. Preparation for Embarkation

A. Planning. Helicopters will often have reduced payloads when operating at higher altitudes. In addition, high temperatures, high humidity, and high Density Altitude will degrade helicopter performance. Consequently, helicopter payloads may change significantly due to both the current and forecasted weather and LZ altitudes.

Marines must have the flexibility to change their embarkation plans based on the varying conditions and helicopter support available. Prior detailed planning by unit commanders will greatly assist in quick helicopter operations.

HELICOPTER	SEA LEVEL	5,000 FT MSL	10,000 FT MSL
UH-1N	6 pax and gear	4 pax and gear	2 pax and gear
CH-46E	16 pax and gear	8 pax and gear	4-6 pax and gear
CH-53E	37 pax and gear	24 pax and gear	18 pax and gear
MV-22	12 pax and gear	10 pax and gear	6-8 pax and gear

- These numbers are estimates only. Actual lift capacity will vary depending on fuel consumption, ordnance on board, time of flight, weather, etc.
-
- The UH-60 should be treated similar to a CH-46 for planning purposes.
-
- The CH-47 should be treated similar to a CH-53 for planning purposes.

B. Personnel. A major hazard to personnel operating around helicopters in cold weather is the wind chill generated by the rotor wash. Exposed skin should be kept to a minimum. If a long wait is expected, warming tents should be erected. At the very least, provide some form of protection from the elements, even if it is only a windbreak.

C. Equipment

1. The team sled should be staged as near the landing point as possible. To prevent the team sled from being moved by the rotor wash, the Marines embarking on the helicopter should lay on top of the sled.

2. Weapons should be in a Condition 4 when embarking the aircraft; magazine removed and bolt to the rear. Muzzles should be pointed down on CH-46, CH-53, and MV-22. Muzzles should be pointed up or outward on UH-1N.

3. No equipment (skis, radio antennas) should be allowed to protrude above the height of a man.

4. Packs should not be worn aboard helicopters due to the restricted movement and the requirement to fasten seat belts before departure. Packs should be staged at the center of the aisle on assault aircraft.

4011. Ahkio Huddle Procedures

The embarkation and debarkation drills (ahkio huddle procedures) described below are designed to get your personnel on and off a helicopter as quickly as possible without severe injury due to rotor blade contact in addition to minimizing exposure to wind chill.

These procedures must be practiced so that they can be performed during periods of extreme weather and reduced visibility.

A. Universal Method. The ahkio huddle has been developed as a universal method for loading and unloading all types of Marine Corps helicopters in a snow covered environment. (Refer to Figure 4011-1) Guides are not recommended or required; however, individuals to be lifted should remove overwhites when conducting ahkio huddles. This contrast in color will provide a recognizable reference for the pilot and aircrew. Having one standard procedure eliminates last minute changes that would be necessary when different models and types of helicopters show up at the LZ. This procedure will:

- Diminish the dangers of troops walking into the helicopter rotor and tail rotor blades.
- Reduces the problems of wind-chill.
- Reduces excessive fuel consumption by the helicopters.
- Decreases the duration of time that a visual and noise signature is presented by a helicopter landing in the snow covered LZ.
- Eliminates much of the delay involved in loading and unloading the helicopter.
-

It is important to remember that:

- The tent team(s) is the basic unit for development of the heliteam.
- All of tent group's equipment necessary for survival against the environment must be on the same aircraft as the personnel.

The helicopter lands with the ahkio huddle at the pilots (right seat) 2 o'clock position (refer to Figure 4011-1). Ahkio huddles(s) must be spaced away from any possible obstacle or far enough apart from the landing sites for other aircraft to allow for the possible drift of the helicopter around the ahkio huddle without conflicting with the obstacles or other helicopters (refer to Figure 4011-2). The helicopter lands next to the huddle of troops that are to be lifted (within the rotor arc). This ensures the best possible reference for the pilot and the greatest safety for the troops. Pilots will execute an immediate wave off if they lose sight of the ahkio huddle. To be performed efficiently even during periods of extreme weather and reduced visibility, these procedures must be practiced continuously.

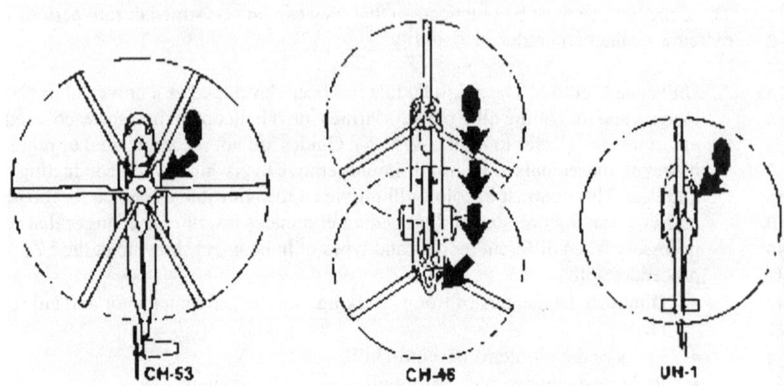

Figure 4011-1: Loading Ahkio Huddles on Marine Helicopters

MV-22 Specific

- Due to the unique flight characteristics of the MV-22, Marines must be aware of the extreme hazard of the engine exhaust from the nacels. The exhaust is over 550 F. (Refer to Figure 4011-3)

- Ahkio teams should position themselves at the pilots 12 o'clock position vice the 2 o'clock position. This will prevent the helicopter from over flying the ahkio team and ensuring engine exhaust does not create a hazard to the troops. (Refer to Figure 4011-4)

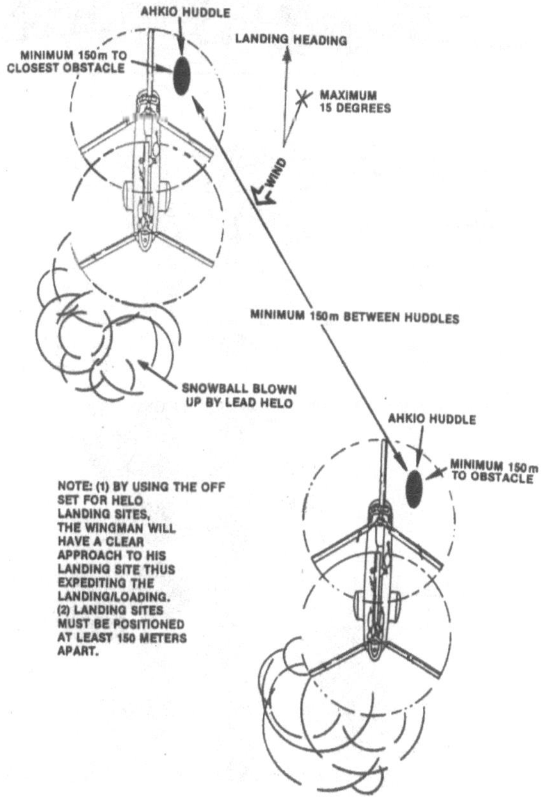

Figure 4011-2: Helicopter in Unprepared Landing Zone

B. Embarkation Procedures. The ahkio huddle is established around the ahkio/group equipment on the landing point. Packs are off; skis and poles are bound together, and snowshoes are attached to packs. Marines group together on top of the equipment, face down, to keep the equipment from blowing away. The helicopter lands beside the ahkio huddle at the pilot's 2 o'clock position. In deep snow, the ramp of the helicopter may not lower enough to embark Marines. Marines must be aware of

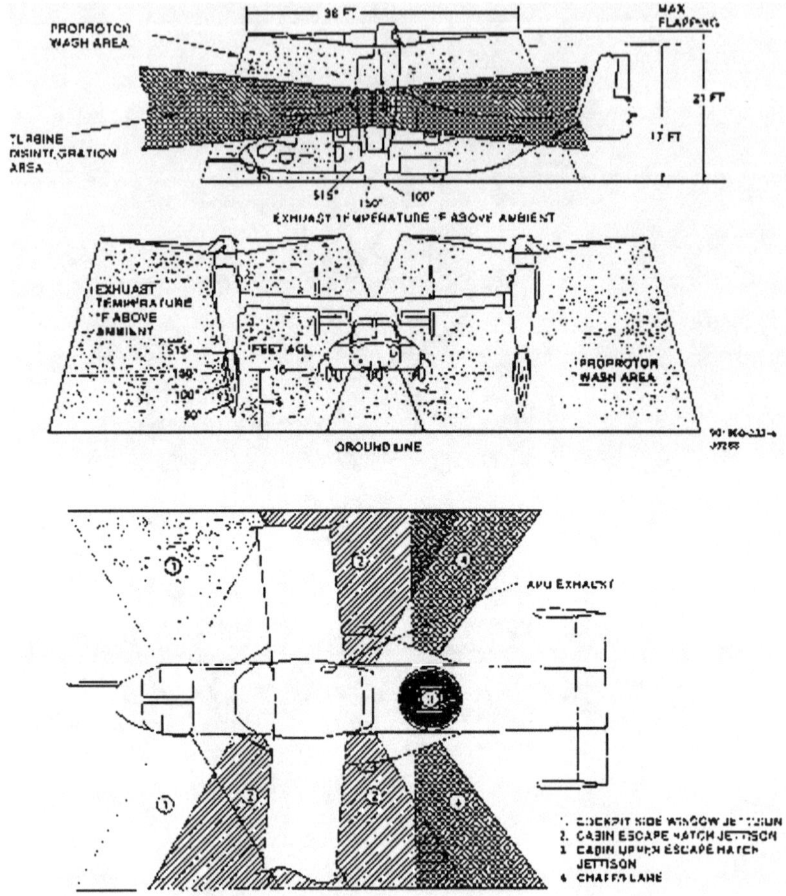

Figure 4011-3: Danger Areas for the V-22

slipping hazards from ice and snow build up in the cargo ramp area. The hydraulically operated ramp may not operate in severe cold temperatures. Marines may have to load via the crew chiefs door on the side of the aircraft.

- Load the helicopter only when directed by the crew chief who will direct the heliteam to load through either the rear cargo ramp or side door.
- The heliteam leader loads first, moves to the front of the helicopter, and coordinates with the pilot.

- Ahkio huddles are located at the 2 o'clock position for all helicopters except MV-22's. Loading is accomplished as depicted in the diagram (refer to Figure 4011-4).
- Designated Marines load equipment near the ramp or exit for they will be the first things offloaded in the new LZ. All other heliteam members enter the aircraft and take their seats. The heliteam leader supervises the loading of the ahkio and any other equipment.
- Snowshoes are strapped to the backpack or team sled.
- Skis and ski poles may be bound together in bundles of four. When loading and unloading, keep skis parallel to the deck at waist level. Once loaded, place skis on the deck of the aircraft beneath the feet.
- As soon as possible after entering the aircraft, each Marine brushes all ice and snow form his uniform and ventilates his clothing to prevent overheating. The crew chief will attempt to maintain the temperature of the helicopter at no more than 40 F. (4 C).

Figure 4011-4: Ahkio huddle positions for MV-22

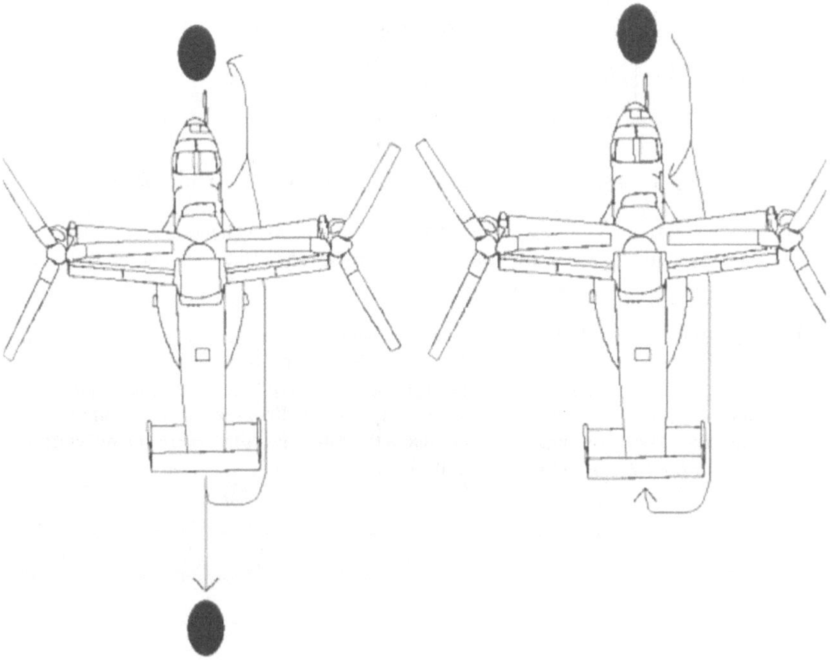

C. Debarkation. As during embarkation, the objective during debarkation is efficiency and safety. (Refer to Figure 4011-5)

- Unload the team sled and other equipment first.
- Then all remaining Marines exit in reverse order of embarkation.
- Establish ahkio huddle. Do not move outside rotor arc until aircraft departs.
- Visibility may be poor when debarking aircraft. Be sure each Marine knows where to go.
- When all equipment and personnel are out of the aircraft and equipment has been secured, the heliteam leader signals the crew chief who indicates to the pilot that it is safe for the helicopter to lift off.

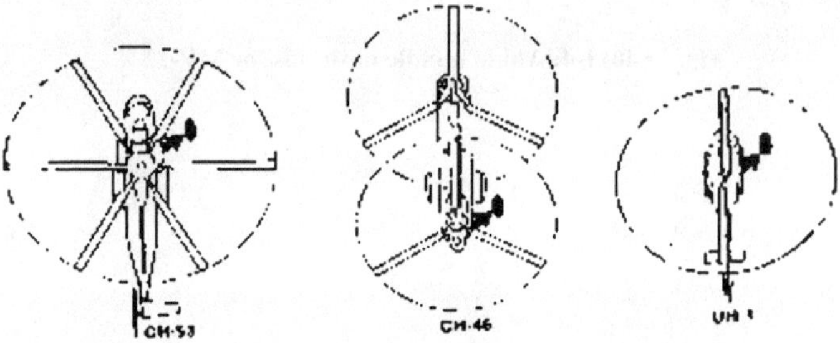

Figure 4011-5: Offloading Procedures for Marine Helicopters

D. Immediate Action in the LZ. The snowball created by the helicopter when it lands can be seen from a considerable distance. Therefore, after the helicopter(s) leave the LZ, move away as quickly as possible.

E. Rotor Clearance. All Marines must realize that helicopters settle into the snow when operating in snow covered LZs. This lowers the distance between the snow surface and the helicopter blades. Using the universal method of helicopter loading (ahkio huddle) eliminates this problem. **Do not approach helicopters from outside the rotor arc. When unloading, do not leave the ahkio huddle before the helicopter debarks the LZ.** (Refer to Figures 4011-6 and 4011-7)

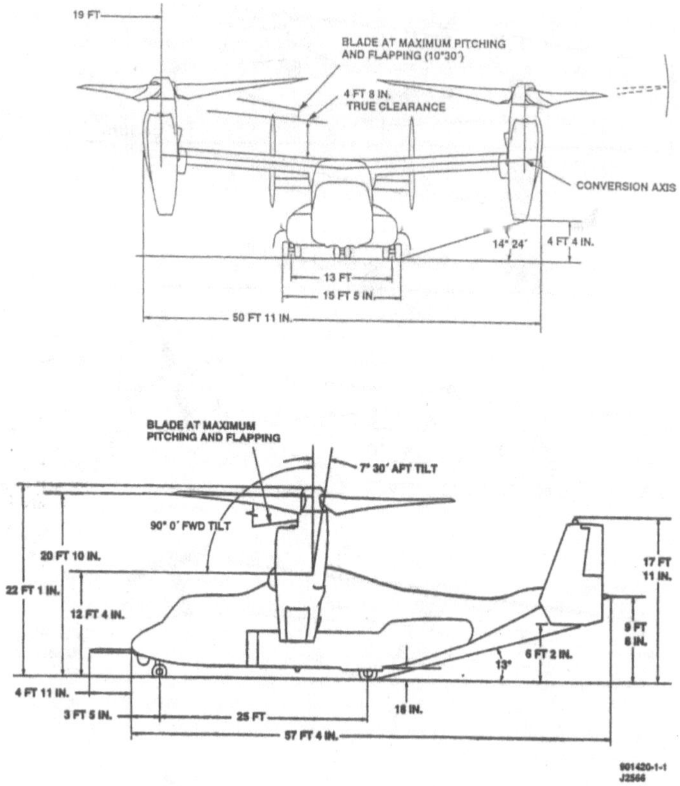

Figure 4011-6: Principal Dimensions and Ground Clearance of V-22

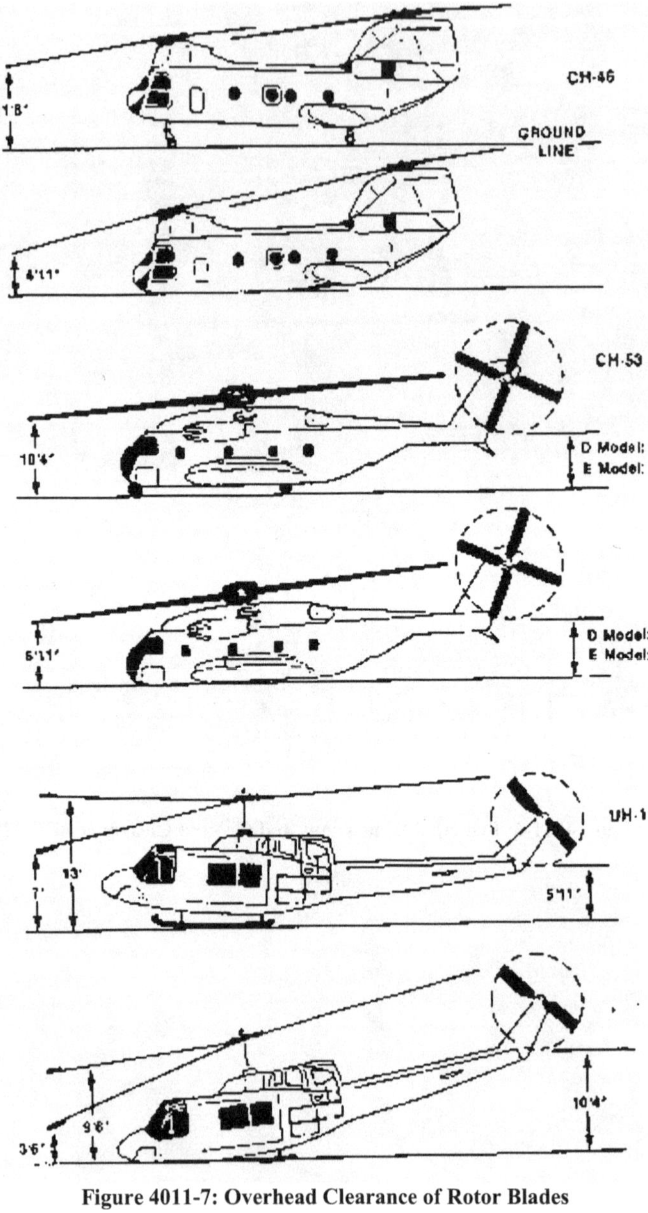

Figure 4011-7: Overhead Clearance of Rotor Blades

4012. External Operations

External operations are not conducted the same in the cold as in other environments. All Marines involved in helicopter external operations must be aware of the unique dangers in pickup and drop off procedures. For detailed information on external lift operations, see FMFRP 5-31, *Multi-Service Helicopter External Air Transport Procedures*.

- **Visibility and Wind-chill**. Blowing snow and ice may obscure pilot flight reference. The rotor wash from hovering helicopters increases wind-chill effect. No skin should be exposed because it will freeze in 30 seconds. Human efficiency is reduced sharply as the temperature drops below -18 C/0 F.
- **Static Electricity**. *Static electrical discharge from helicopters can be fatal to ground personnel.* Low arctic humidity causes static buildup to be much stronger than in warmer climates. Ground personnel must ground hoists and pendants before touching. Insert grounding rods securely through the snow into solid ground. Hookup men must wear high voltage anti-static gloves.
- **Protective Equipment**. Marines receiving the load should wear goggles, a hard hat secured with chinstraps, and high voltage anti-static gloves.
- **Loose Gear**. Loose gear within 100 ft. of the pickup and drop-off points must be secured.
- **Loose-Fitting Clothing**. Do not wear loose-fitting clothing. It will flap in the rotor wash and snag in hoist lines.

A. Pickup LZ Procedures.

1. Snow must be compacted as much as possible to lessen the snowball created by rotor wash.

2. Helicopter support team (HST) directors must remove overwhites, put on international orange road guard vests, and use signal paddles.

3. HST hookup crews for the next load must remove overwhites and put on international orange road guard vests.

4. Remaining Marines in the LZ will wear overwhites without international orange road guard vests to prevent confusion.

5. Several approaches at a higher than normal hover altitude may be required as the pilot attempts to blow away snow and ice before external lift operations start.

6. The snowball created by rotor wash can create a signature that can be observed for miles. Hovering over a pickup or drop-off point to clear the LZ of blowing snow will have to be tactically evaluated in an active combat zone.

7. If the helicopter is using longer pendants than normal, the HST director must increase his distance form the load so that the pilot can see him.

8. The helicopter must face into the wind when picking up a load. Downwind pickups with the snowball in front of the helicopter will cause the pilot to lose sight of the HST director. (The helicopters nose will be too high for the pilot to see the HST director (refer to Figure 4007-1).

9. Visual or radio communication must be maintained at all times between the helicopter and the ground crew.

B. Drop-off LZ Procedures. Basically, the same procedures for pickup apply to drop-off procedures. However, only the designated HST director will have signal paddles and wear an international orange road guard vest. Overwhites are not worn. The distance of the HST director is increased if longer pendants are used.

1. To mark the drop point, anything that contrasts with the snow, will not blow away, and is visible to the crew chief and pilot, will do. Use smoke grenades during the helicopter's approach to identify the drop point. However, as the helicopter comes into a hover and needs a reference point, the grenade usually goes out or can obscure the landing site. As an alternate, use:
   - Pine boughs.
   - Dye markers in letters or geometric shapes.
   - Mess hall food coloring poured on the snow.
   - Snow filled dark-colored trash bags. These must be anchored down.

2. Do not expect the HST director to provide final guidance to the drop point. The pilot will be directed by his crew chief who will have the drop point in sight. The crew chief will assist in the final positioning of the helicopter. (Less than 20 meters will degrade the quality of the hover and increase the chances of the helicopter being engulfed in a snowball.) **Once the helicopter's hover is degraded or the pilot losses sight of the ground, he will either wave-off or drop his external load!**

3. When lifting artillery into areas covered by deep snow, advance parties must precede delivery to prepare the gun pits. Pits must be large enough to maneuver the trails inside the pit, and deep enough to prevent the howitzer from sinking down into the snow. The general azimuth where the artillery should be pointed should be marked on the snow with a dark line with an arrow at the bottom of the pit. The helicopter crew chief can then visually align the barrel to desired azimuth before releasing the load.

4. As is pickup procedures, communication must be maintained at all times between the helicopter and the ground crew.

4013. Fixed Wing Operations

A. Deicing procedures

1. Deicing Fluid**.** Deicing fluid is effective on frosted or ice-covered surfaces, but is less effective on snow. Snow and deicer form a mixture, which is hard to remove. Always remove loose snow with a broom or brush before applying deicer but not on the canopy! When icing is expected or when planning to taxi or tow aircraft over slush-covered runways, apply undiluted and unheated deicer fluid. Use MIL-A-8243, Anti-icing and Deicing/Defrosting Fluid. Use a sprayer, soft brush, or broom. **Deicing Fluid is toxic. Do not breathe the fumes or get fluid on the skin or in the eyes.**

2. Bearings. Do not get deicing fluids in the bearings. Solvents may dilute the grease. Water may freeze.

3. Hardened Ice. Remove hardened ice with diluted deicer heated to 180 degree F to 200 F. Apply in a solid stream to flood away the ice. Then spray deicer to prevent re-freezing. Soft cotton cloths or bedding blankets can be saturated with dicing fluid, squeezed nearly drip free, and laid on the aircraft. Pour more fluid on the covering until the ice melts. Unless the aircraft is to be flown immediately, wipe it off and wash with water.

4. Use Before Ice Forms. Deicing fluid will prevent ice if applied before icing conditions occur. Snow and ice will dilute the fluid and wash it off. If rain or snow is expected, apply the dicer and cover the aircraft.

5. Use Before Ground Taxiing. Spray or brush the deicer on the aircraft's underside especially if on slush-covered runways.

6. Uncovered Aircraft. To protect uncovered aircraft at night when rain or snow is not expected, coat with full-strength, unheated deicing fluid. The aircraft should be ice-free in the morning.

7. Fog. Ice forms rapidly in the fog. Therefore, within a short time after deicing, you'll probably have to perform a second deicing before flying the aircraft.

8. Condensation/Icing. If the aircraft is parked in warm areas, make every effort to dry the aircraft before moving it into the cold air. Condensation will occur. Deicing will be required.

B. Preflight

1. Hydraulics. Hydraulic leaks may occur because fittings contract and packings shrink or become distorted. Ice crystals in hydraulic fluid may cut seal materials. When the temperature is near –40 F, hydraulic fluid is likely to lead from the bulkhead universal couplings, running seals, and long-stroke actuators. Check for leaks from actuating cylinders. Unlock cylinders on the nose gear and main landing gear, outer wing cylinders, speed brake cylinders, wing fold cylinders, and accumulators. Before replacing a leaking hydraulic mechanism, move the aircraft into a warm hanger. See

if the leak stops or apply heat to the mechanism for 1 hour and check again. Three of four cycles of a hydraulic mechanism will usually warm it up enough to stop minor leaks. If replacement is necessary, finish the work in the warm hanger for proper seating of "O" rings and tubing connectors. The cold makes mechanical and hydraulic equipment sluggish. ***If aircraft engines are cranked repeatedly, you may damage the transfer gearbox of the constant speed drive.***

2. Snow or Ice Removal. Remove all pressure-sensitive tape from wing and flap junctures. Remove all protective covers by lifting rather than sliding. **You could scratch the canopy. Remove loose snow with a broom or brush but not on the canopy. Do not use tools or scrapers that may damage the airframe**. The transparent acrylic plastic of the canopy can easily be scratched or marred. Do not clean the canopy with a broom or with solvents such as deicing fluids.

3. Doors and Panels. Doors and panels may freeze shut during cold weather. To break the ice seal, apply deicing fluid or heat to the edges of the door or panel. A panel or stress frame stored in a warm area may not fit properly if you try to immediately install it in a cold aircraft. Wait a few minutes before installing. Let the metal contract.

4. Exposed Mechanisms. Check that the following are free of snow, ice, and dirt:

Nose landing gear strut.
Main landing gear strut

5. Water in Cockpits. Water dripping into cockpits may short switches or damage instruments. When the front canopy is open, make every effort to prevent water from dripping off the left sill onto the engine master switches. If water accumulates in the cockpit, open the drains in the nose of the landing gear well. To avoid loss of cabin pressure later in flight, make sure that the drains are resealed as soon as the water drains.

6. Oxygen Mask. If the oxygen mask is not fastened, keep it away from the face. This prevents the breath from freezing in the inhalation valve. In temperatures below 0 F, it may be hard to connect the oxygen mask to the T-connector due to a stiff "O" ring. If it is hard to connect the mask nose to the T-connector, apply heat to the connector.

7. Windshields. Operate the foot heat and defog control levers to increase heat to the windshields. If the outside of the windshield is iced, operate the rain removal system.

8. Strap In. Assistance from the plane captain may be required to strap in and remove the seat pins, due to the bulk of cold weather flying clothes.

9. Tires. Make sure that all tires are always in good condition and are inflated to recommended pressure. Inflate tires with dry nitrogen rather than air. **Do not mix**

air with Class 2 nitrogen.

C. Aircraft Turn up. Do not attempt to cycle the flight controls until signaled by the plane captain. If any ice or snow is present upon start, abrupt movements of the stick may cause control rod breakage.

D. Ground Taxiing and Towing. When ground taxiing or towing aircraft on snow, ice or slush, go very slow! Aircraft thrust is much greater at low temperatures. Allow greater distances between aircraft and other objects even when taxiing on dry surfaces. Taxi at 10 knots or less. Turn gradually. Do not jab the brakes. To prevent freezing of the brakes and icing of the fuselage's underside, do not taxi through snow. Use the brakes intermittently to avoid filling the tire treads with snow or ice. In heavy snow, mud, or other rough terrain, do not push or pull aircraft with only a tow bar. Taxiing at minimum power will prevent wheels and brakes from accumulating mud or slush that can freeze in flight. Avoid taxiing in loose snow. Moisture may be ingested through the engine auxiliary air doors.

1. Slipping. Ice at 30 F is twice as slippery as ice at 0 F. Be extremely careful when taxiing on icy surfaces when the temperature is near 32 F.

2. Run-Ups. If engine run-ups are performed, select an ice-free area. Remember the aircraft will skid with the increased thrust. If taxi speeds exceed 10 knots, collisions can easily occur on runways covered with snow, ice, or slush.

3. Taxi Interval. Increase taxiing intervals to create a safe stopping distance and to prevent icing of the aircraft surfaces by snow and ice melted by the jet blast of the preceding aircraft at least 300 feet behind the aircraft in front. This spacing will avoid collisions and will reduce icing from the melted snow and ice thrown by the jet blast of the preceding aircraft. A good engine run-up is probably not possible without throwing snow and ice. Engine exhaust may crate a cloud and reduce visibility of following aircraft.

4. Canopy Frosting. Snow, ice, and frost can obscure visibility through canopies and windshields. If the aircraft air-conditioning system is not operated properly, condensation can cause canopy fogging and frosting inside the cockpit.

5. Aircraft Preheating. During cold weather operations, pre-heating electronic bays and cockpits can reduce warm-up time. When preheating cockpits, leave the canopy open about 1 inch from the windshield. Do not apply heat directly to the windshield or canopy. Remove accumulated snow before applying heat. Make sure melting snow or ice does not run into flight controls. When using heaters, make sure adequate fire-fighting equipment is readily available.

6. Landing Gear. To avoid damaging the landing gear during sub-zero temperatures, do not steer the nose wheel while the aircraft is parked. Taxi carefully over rough or rutted ice and hardened snow. Damage to the landing gear could result from

stiff lubricant in the struts and wheel bearings. Avoid excessive movements of the landing gear shock struts. You can damage the seals. Avoid excessive power or fast starts with a tow truck. If aircraft tires are frozen to the ramp or if the wheels hit ice chunks, high breakaway loads may damage the landing gear.

7. Nose Wheel Struts. Attach cables or chains to the tie down rings on the main wheel struts to avoid damage.

E. Takeoff. When operating from runways that are covered with excessive water, snow, or slush, high-speed aborts may result in engine flameout due to precipitation ingestion. Flameout probability is highest when throttles are chopped from after-burner to idle at speeds above 100 knots. Leave the landing gear down for about 1 minute to break off snow and ice.

F. In-flight Ice Buildup. Freezing rain can cause a rapid buildup of ice. Treat buildup as a full-scale emergency requiring a change in altitude and airspeed. If in-flight freezing within the longitudinal control system happens, excessive force may be required to move the control stick. Normal airplane control is available but requires higher stick force inputs.

1. Fuselage Icing. Icing is likely when the relative humidity is high and the outside air temperature is between 10 F and 32 F. Taxing through snow and slush will build up ice on the fuselage's underside.

2. Instrument Sensor Icing. Ice impact damage or ice buildup on external sensors can cause incorrect airspeed and altitude readings. Trouble may occur in the static pressure correction, total temperature sensor, angle of attack, and longitudinal feel trim systems.

3. Engine Icing. If the aircraft anti-icing system is not used, engine icing may occur when the temperature is between 10 F and 32 F. However, the cooling effect at low airspeeds may cause engine inlet icing at temperatures as high as 41 f: i.e., when the dew point is within 7 F of the outside air temperature.

G. Inadvertent Instrument Meteorological Conditions. Inadvertent entry in instrument meteorological conditions (IMC) usually poses little problem. Climbing to a higher altitude will normally alleviate the problem. Exercise caution to remain aware of the highest terrain in the area. If VMC conditions cannot be reached, stay on the gauges. **Aviate, Navigate, and Communicate!**

H. Landing

1. Use anti-skid, if equipped.

2. As soon as practical after landing roll, place flaps in the full up position.

3. When the temperature is below or forecasted to be below 32 F with heavy rain, aircraft may be parked with flaps down and wings spread.

I. Post-flight

1. Canopies. To prevent cracking or frosting of canopies and windshields in warm aircraft which are to remain in the cold, leave canopies open about 1 inch from the windshield for 5 minutes, weather permitting. This allows the cockpit temperature to approach the outside air temperature

2. Fuel Tanks and Fuel Cells. Condensation causes water to accumulate in the fuel tanks and fuel cells, especially if they are not kept full. Refuel aircraft as soon as possible after landing. If water freezes, it may clog filters, fuel lines, and valves.

3. Covers. To avoid damage and prevent freezing of covers to the aircraft, make sure that contracting surfaces of the covers and aircraft are free of dirt, grit, grease, ice, snow, slush, and water. Protect the aircraft by installing covers over the following areas:

Canopy and radome.
Rain removal system
Engine air ducts.
Pitot tube.
Engine afterburners.
Angle of attack probe.
Total temperature sensor.
Wheel folds.

4014. Effects of the Cold on the Functional Areas of Marine Aviation

A. Air Reconnaissance. The cold increases reliance on air reconnaissance over other information gathering assets. Cold and snow less affect air reconnaissance assets (once airborne). This increased reliance on air reconnaissance to gather information will require an increase in air assets and support equipment/personnel. The snow affects air reconnaissance as it masks relief and definition normally found in the landscape, therefore decreasing the effectiveness of visual observation. However, the snow enhances the capabilities of infrared and thermal imagery equipment.

B. Electronic Warfare. Marine tactical electronic warfare (EW) squadrons are the only specialized Marine airborne EW units. Their primary mission is to provide EW in support of Fleet Marine Force (FMF) operations. The airborne EW platform is classified as an all weather platform. However, like other aircraft, it cannot sustain ice accumulation. Electronic pods and antennas cannot function under icing conditions. Weather conditions must always be considered when developing EW packages.

C. Assault Support. Helicopter operations in soft or loose snow will increase signatures making nap of the earth (NOE) routes easy to identify. Movement into loose snow-covered LZ's will result in identifying signatures, which may result in enemy artillery fires. These signatures which may result in enemy artillery fires but can also be used to conduct deception operations. Take special efforts when constructing LZ's to prevent or reduce identifying signatures. All movement times will significantly increase in extreme cold temperatures.

D. Offensive Air Support. Flight operations may need to be conducted at night or in reduced visibility where enemy observations are significantly reduced. Up to 90 days of continuous darkness are found in some cold weather operating areas. The cold weather areas will present difficulty with target marking in snow-covered terrain or with laser designation.

E. Anti-air Warfare. The basic principles of passive and active air defense remain the same in cold weather operations. The MAGTF's anti-air warfare (AAW) focus of effort must consider the scarcity and importance of ports, airfields, main supply routes, and supply dumps, and that protecting these facilities will assume greater importance.

F. Control of Aircraft and Missiles. Control of aircraft and missiles is accomplished throughout he Marine air command and control system (MACCS). The MACCS agencies are:

Tactical air command center (TACC)
Tactical air operations center (TAOC)
Direct air support center (DASC)
Marine air traffic control squadron detachments.
Air support radar teams
Marine wing communication squadron detachments

G. The effects of cold weather on the MACCS is primarily concerned with:

1. Operational Area Implications. Site selection of various MACCS agencies and units is critical of various MACCS agencies and units are critical for mission accomplishment. Radar coverage, weapons range, and communication connectivity are all adversely affected by mountainous terrain. Access to hilltops, may require extensive helicopter support. High winds, ice, and snow will affect antenna operation and other equipment functions, and can even terminate operations. Detection of the low altitude terrain-masked threat becomes harder. Displacements are harder and require longer time to execute.

2. Support. Support problems increase significantly in cold weather. Snow removal for site locations is required even in moderate snowfall. Alternate and supplementary positions must be developed. Anchoring antennas and equipment is more difficult because of characteristic high winds, rock-frozen

soil, and ice. The MACCS usually depends on motor transport (MT) assets for tactical mobility. Therefore, all of the inherent problems associated with cold weather MT are present as well. Tactical displacement by helicopter is an alternative although difficult, time-consuming, and presents difficult signature problems. Equipment and cable placement require dunnage or hay to prevent them from freezing into the ground. Displacement times are increased two to five times that of a normal operation. Host nation support for snow removal, drilling, material handling, and engineering support must be considered.

3. Communications. The MACCS cannot operate effectively without communications. Use relay sites to provide agency connectivity. Extensive use of host nations tactical circuits, phones, and commercial equipment is often required for interoperability with national command and control facilities and as redundant pathways. It is critical that subordinate leaders understand the commander's intent so the MACCS nodes or agencies can conduct autonomous operations when communications are degraded or destroyed. Problems that must be solved include equipment installation and grounding and safety.

4. Personnel/Equipment. Each MACCS agency has a significant amount of personnel and equipment. The cold significantly drains human energy and will cause injuries or casualties. The limited number of organic prime movers available to each of these agencies makes mobility difficult. Cold weather compounds this problem since standard tactical vehicles may be unsuitable. Therefore, support of MACCS agencies must be considered when the MAGTF allocates over the snow vehicles.

CHAPTER 5

COMBAT SERVICE SUPPORT OPERATIONS

5001. General Effects of Cold Weather on Combat Service Support

A. The Enemy and The Environment. Logistical support is more critical in the cold than in any other environment. The enemy and the environment confront the commander. Underestimating either of these threats can result in failure. History is full of examples of forces that neglected or disregarded their combat service support (CSS) planning and needs and failed because of the effects of the environment. Cold weather combat has been decided more by the environment than by combat with the enemy. Neglect of CSS in the past and in the future will produce significant combat losses. Most of these critical losses are of the non-battle variety.

B. Staff Planning Functions. The basic concepts and principles of logistical support will not change from those described in FMFM 4-1, Combat Service Support Operations.

1. Time and Space. The influence on time and space by the cold weather environment must always be considered. The cold, snow, ice, transition weather, and their effects on mobility and human tasks must be considered combat multipliers.
2. Concept of Supportability. Situational assessments and the various staff estimates of support evolve into considerations that, in turn, translate into requirements. Failure to anticipate logistical needs to conduct and execute detailed planning for CSS means disaster. The logistics officer's concept of support may affect the options that the commander will have when making command decisions. The commander will have to determine CSS requirements, establish priorities, and allocate resources with greater precision than when operating in other environments. He may find that his commander's guidance must be modified because of logistical constraints.
3. Flexibility and Redundancy. Flexibility in CSS operations is essential. Both supported and supporting CSS planners must always consider alternate means of fulfilling identified requirements. Redundancy will be necessary to ensure responsiveness. In an amphibious operation, more seabasing of supplies than normal may be necessary. This may be impacted by the vehicles to be used and the physical condition of bays and ports. The nature of cold weather operations dictates redundancy over economy in CSS.
4. Light/Darkness. The reduced hours of daylight consistent with arctic or sub-arctic winter operations may dictate the need for artificial lighting. The provision of the additional lighting must be planned for in the CSS concept of operations. Conversely, the extended periods of daylight in transition and summer periods will require important operational adjustments.

C. Preparing for Embarkation. Responsibility for tasks of preparing critical items of supply and pieces of equipment must be clearly delineated.

1. MAGTF Responsibilities. A special organization unique to cold weather deployments needs to be created by the MAGTF to ensure that the specialized shipping, storage, and handling of logistics are accomplished. This organization is the movement coordination center (MCC). It will consist of maintenance management and medical personnel task organized from CSS assets. Established in the MAGTF rear, it is responsible for assuring that each piece of equipment, ordnance, and supply has been properly prepared for cold weather operations before embarkation. The MCC:

    - Performs limited technical inspections (LTIS) on all vehicles, equipment, and supplies.
    - Prepares vehicles with correct fuels, lubricants, and fully charged batteries.
    - Provides additional petroleum, oil, and lubricants and lubricant blocks for continued operation.
    - Ensures the items in the AMAL are in marked containers, stored, and shipped, properly protected from the elements.

2. CSSE Responsibilities. In the MAGTF rear, the CSSE will provide personnel to establish and implement the MCC. In country, the offload, startup, and repair of damaged or malfunctioning vehicles are the CSSE's responsibility once the equipment arrives.
3. Unit Responsibilities. Preparation and pre-operation LTIs are unit responsibilities. Each major subordinate element has responsibility for embarking, blocking, bracing, and dogging down its equipment and supplies on roll-on/roll-off rail or air transport at point of embarkation.

D. Time and Distance. Distance will be measured in time rather than mileage. Time and distance will be at least doubled by the environment. It may take up to five times as long to accomplish some CSS tasks. The environment is extremely demanding of manpower. Trafficability will be so unpredictable that every possible alternative must be considered. Assets will be limited. Roads will be limited if existing at all. Cross-country movement for normal transportation assets may be impossible. The environment will create a reliance on the push method of logistical support versus the reactive pull system created by rapid requests. Needs must be anticipated. This will require preplanned supply and resupply operations, depending to a great degree on helicopterborne or aerial delivery methods that are greatly dependent upon the weather. At all times, support will be subject to disruption by the weather or other environmental phenomena. Avalanches, sudden thaws (mud and flooding), or local storms may bring operations to a halt.

E. Wet Cold. Transition periods during cold weather operations are particularly dangerous. History indicates that armies will fight from village to village in the cold to protect their soldiers from the elements. Wet cold is particularly dangerous and has the potential to kill. Hypothermia weakens individuals and renders them helpless. Indigenous forces historically have shown a decided advantage due to the support of their native populations.

1. MAGTF Responsibilities. The G-4/S-4 must anticipate well in advance, problems like wet sleeping bags that will essentially be unserviceable if not dried. The G-4/S-4 must be prepared with viable solutions to problems presented by wet old/transition weather. Host nation resources may need to be called on when requirements exceed organic capabilities.
2. CSSE Responsibilities. The CSSE must be prepared to provide CSS to all elements in the field under all conditions. Needs must be anticipated well in advance. During wet cold conditions and transition periods, roads undoubtedly may be closed or nonexistent, and helicopter support limited. The CSSE must be prepared to implement solutions to problems identified by the MAGTF.
3. Unit Responsibilities. In wet cold conditions, uniforms and equipment will get wet. Commanders must also have the foresight to provide warming tents and stoves if their units are to survive. These assets should be in the unit logistics train. In anticipation of wet cold and transition periods, it may be necessary to prestage these items forward. For this concept to work in combat, it must be exercised in training.

F. Realistic Training. Training operations in cold weather carry some, but not all, of the same environmental risks and problems as combat operations. Training must be realistic. CSSEs must train to operate and fight in the cold before deployment. Every effort must be made to simulate the same types and quantities of loads that will be encountered in combat. Simulated movement of supplies, casualties, and evacuation of damaged equipment must be included in all exercises.

G. Information Systems Management Officer Functions. Computers are susceptible to temperature and humidity changes. The computers used include the Green Machine at the unit level and the deployable force automated service center computers. They may be subject to failure under cold conditions. They must be used, stored, and transported in heated spaces. Microscopic gaps in electrical and magnetic connections are prone to short-circuiting from moisture or ice accumulation. Disks will be brittle when exposed to the cold. Operating systems must not be activated until a heated environment has been established and hardware has come up to temperature.

5002. Communication Considerations.

Communicating in a cold weather, mountainous environment presents unique challenges that must be identified and overcome. Weather, geography, and altitude are all factors that can effect equipment, personnel, and communications organization. The key to success in combat is reliable, secure, rapid, and flexible communications. This is especially necessary in the cold where the units face problems of both survival and mission accomplishment. Commanders must understand cold's effects on their communications systems and their personnel, and know the procedures to counteract these effects.

A. Planning Considerations

1. Planning Functions. Communications tasks are more difficult in the cold. The toll on personnel is tremendous. The communications officer must conduct reconnaissance, plan frequencies within ranges that will work, and request additional personnel. Commanders must plan for:
    - Communications equipment
    - Communications maintenance and supplies
    - Safety
    - The equipment load communications personnel must carry in addition to their required personal equipment.
    - Communications plans
    - Additional personnel and equipment needed to man retransmission sites and to conduct mountain-picketing operations.
    - Communications system configurations

2. Site Selection
    a. Impact of the cold. The biggest impact of the cold on site selection will be the tendency to establish antenna farms near or within the regimental or battalion command post (CP) perimeter. At company level, the radios are in the CP anyway, so this is not a consideration. Locating the antenna farms near or within the CP will reduce wire and cable problems, but will render the CP more vulnerable to direction finding. This will dictate that CP sites be selected in good positions from which communications can be conducted and still protected from the enemy direct fires.
    b. Displacement. Communications Officers must constantly practice displacement in the day and night. Commanders should consider deceptive movements and using phony CPs. Communications variant vehicles can be used as mobile CPs.
    c. Tree / Snow Lines. Tree/snow lines exist in many cold weather contingency areas. Communicators must always take this into account when installing antenna farms and retransmission sites. If possible, these sites should be established below the tree/snow line where they can be camouflaged.

3. Redundancy. Communications in low temperatures and high latitudes is extremely difficult. Operational success may depend on reliable communications because of dispersion of the unit. Commanders and communications officers must plan on using all available communications assets. Backup capabilities must always be planned for and used. These assets should include wire, VHF, and host nation capabilities; AAVs, LAVs, and BV-202/BV206s; and messengers/carriers using vehicles, motorcycles, snowmobiles, skis, or snowshoes.
4. Communications Support. The concept of communications support does not change in cold weather. All elements of the landing force afloat are provided communications support by the Naval Commander and the amphibious ship upon which they are embarked. When ashore, the MAGTF command element (CE) receives communications support from the SRI detachment (primarily the supporting communications battalion). MAGTF subordinate elements use organic communications assets from their parent commands to support internal communications requirements as depicted below:

    - MarDiv Communications Company, HQ Bn
    - MAW Marine Wing Communications Squadron, Marine Air Control group
    - FSSG Communications Company, H&S Bn

5. Extended operations ashore will require the GCE, ACE, and CSSE to communicate with the MAGTF CE, and possibly with other joint and DOD agencies (such as Defense Communications Agency). These communications requirements will normally be satisfied with assets from the supporting Communications Battalion.
6. Planning Responsibilities. The MAGTF Communications/Information Systems Officer) is responsible for the overall planning, coordination, and supervision of the landing force's communications system. Cold weather operations will require that the G-6/S-6 at all levels of command consider the need for additional equipment and personnel to sustain effective communications in the harsh environment. Augmentation of communications assets from the supporting communications element will be required to support communications requirements for sustained operations.
7. Repair of Communication Assets. Repair of communications assets is always a continual concern but is intensified by the harsh effects of the cold weather environment. The CSSE is responsible for the overall planning, coordination, and supervision of the 3rd/4th echelon communications-electronics maintenance for the MAGTF. Planning must consider the need for additional personnel, extra supply of repair parts, and heated maintenance facilities necessary to sustain effective communications-electronics maintenance support.
8. Wireless (Radio) Communication. Radios are the most common means of communicating. They are subject to many problems in the cold. The two major problems are reduced battery power and increased equipment failure. Other problems include increased incidence of ground reflection on radio waves, polar

atmospheric conditions, and radio remoting. All radios should be in good operating condition before cold weather operations are initiated. Radios and ancillary components must be kept dry and sheltered from the cold weather. Constructing shelter (snow caves, igloos, shelter halves/tent sheets) around the radios will protect them from the environment and will raise the ambient temperature around the radios. Radios left on may operate more effectively due to the heat generated internally. Unpowered radios and batteries stored in a sub-zero environment may be difficult to activate and operate less effectively.

a. Battery Power. Batteries of all types give less power at low temperatures. The conventional dry cell battery loses efficiency very rapidly as temperatures fall. Dry batteries should be stored at a temperature above 10 degrees F and gently warmed, either with body or vehicle heat, before use. They should not be exposed to extreme cold until needed. During use, they should be kept as warm as possible. Man-packed radios and batteries should be taken into shelters overnight. NOTE: Continually moving radios from warm to cold weather environments will cause condensation inside radio equipment. Once taken into a warm environment, radios may have to be stored in a *hot box* or radio equipment may be stored in a cold shelter with batteries removed and stored in a warm shelter. To confront effects of cold weather radio operators should:

- Always use cold weather batteries. (Refer to Fig. 5-1)
- Never place the radio in the snow. If snow covers the pressure release cover, ice may form which can restrict the radio battery box from air exchange.
- Rotate batteries at least every 4 hours.
- Carry one spare set of batteries in a parka or trouser pocket between the body and outside protective clothing. Body heat keeps the batteries warm.
- Keep log entries when batteries are changed.
- Store spare batteries inside heated shelters.
- Insulate batteries from snow or ice contacts with whatever means are available.

b. Resupply. Batteries require a one-for-one exchange.

c. Lithium Batteries. Lithium batteries are superior to magnesium batteries in the cold. They are lighter and last longer. Batteries perform best and last longer when kept cool, rather than cold or freezing temperatures. Optimum storage temperatures are +35 F or slightly colder. They should not be allowed to freeze. The plastic female connectors on BA-5590 batteries become particularly brittle when cold and must be handled gently.

d. Cold Weather Batteries. Some batteries have cold weather counterparts. The BA-3030 is the cold weather replacement for a BA-30. A close watch is necessary to ensure that BA-30's are not substituted for BA-3030's (figure 5-1 lists cold weather batteries).

9. Material Failure. Flexible cables and some metal and plastic parts become brittle at low temperatures. All cables and rubber parts should be treated with silicone

lubricant to prevent them from becoming brittle and cracking. Rough handling can easily break power connections and cables. Before they are connected, they should be warmed so that they can easily be manipulated without damage. When temperatures are below 10 F, radio equipment materials become brittle. Some are very susceptible to breakage. The more common problems affect:

NSN	**NOMENCLATURE**	
6135-00-930-OO30	BA-3030	
6153-00-935-2587	BA-3030	
6153-00-935-2587	BA-3058	
6153-00-935-5001	BA-3042	
6135-00-850-3177	BA-90	
6135-00-801-3493	BA-1372	
6135-00-050-3280	BA-200	
6135-00-464-7584	BA-3553	*
6135-01-034-2239	BA-5598	*
6135-01-036-3495	BA-5590	*
* Also used in temperate climates.		

Figure 5-1: Batteries for Cold Weather

a. Antennae. Antennae may be difficult to erect in deep, soft snow and on frozen ground. They are likely to become iced-up and are susceptible to a phenomenon called precipitation static. Antennae, and particularly the support wires, should be jarred frequently to dislodge any ice. Wire antennae should be erected so that the wire is attached to one post by string, which is of a thickness that will break before the antennae, under the weight of ice buildup. An antennae can be given additional height in an area where there are saplings by bending over one of the saplings, attached the antennae, and then releasing the sapling.
   - AT-2171A (10-Foot Whip). It is difficult to ski or move in any type of vegetation. Remove the AT-271A while skiing or moving through vegetation. Radio operators should carry a spare AT-271A.
   - OE-254. Coaxial cable, connectors, and antennae elements need a thin coat of silicone lubricant spread on their components.

b. Guys. In snow anchor guys by using the deadman technique or tie off guys to stationary objects, such as trees, when available. Do not use fewer guys than the antennae technical manual calls for. Icing and high winds will tax guy lines heavily. Ice also reduces antennae radiating ranges and poses a safety hazard. Inspect and clean off ice regularly. Keep ceramic insulator bowls dry and free of ice and handle them carefully. Cold makes the ceramic brittle. Keep antennae whip and mast sections dry and free of dirt. Apply a light film of silicone lubricant to each joint. Silicone comes in 2-oz tubes, NSN 6850-00-177-5094, and 8-oz tubes, NSN 6850-00-880-7616. Be gentle when coiling or uncoiling antennae coaxial cables. Cold makes the cable brittle. If possible, warm cables prior to installation.

c. Couplers. Antennae couplers on GRC-193/MRC-139 radios do not operate in cold weather. These couplers freeze up making tuning impossible. Shield these couplers with insulation; windbreak the area; use plastic and tape, etc., to protect the antennae couplers from the weather.

d. Expedient Antennas. Installing standard antennae systems becomes more difficult in cold weather operations. The effects of temperature below the freezing point, particularly sub-zero (F) temperatures, make antennae lead wire/cable with rubber or plastic sheathing brittle and susceptible to breakage or even shattering. Coaxial antennae lead cable is especially vulnerable to sub-zero temperatures and is very difficult to install or move when it freezes and is not flexible. Antennae elements and connector leads must be kept free of moisture to prevent equipment damage resulting from the expansion of freezing water. It is very important to make sure all antennae elements are in good operating condition and that connectors and wire/cable are treated with a cold weather lubricant before cold weather operations. Large erectable antennas, such as the OE-254 or AS-2259, require antennae guy wires to hold them in place. Frozen ground and snow covered ground make placement of guy wire anchors more difficult and often require auger-type guy stakes made of steel, trees, other natural objects, or the building of deadmen to which the guy wires may be attached. Consider using field expedient antenna systems with field communications wire (WD-1/TT). This wire is less vulnerable to breakage and is much easier to install during sub-zero temperatures. For highly mobile units the use of expedient antenna systems which do not have to be retrieved and are easily erected with expendable communications wire. The use of field expedient antennae to support HF/VHF tactical communications is a critical aspect of C2 for the commander. Reference material on field expedient antennae is available from the Communications Officer School, Marine Air Ground Training and Education Center, Marine Corps Combat Development Command, Quantico, VA.

- Remotes (AN/GRA 39). Cables and connectors need a thin coat of lubricant.
- Handsets (H-250 or H-189). The handset cable and connector must have a thin coat of silicone lubricant. The press-to-talk button is subject to sticking from freezing. Wrap the plastic bag from the BA-4386 around the handset to prevent the press-to-talk button from freezing. Radio operators should carry spare handsets. A moisture cover; i.e., a plastic bag, for handsets will prevent moisture from freezing in the microphone.

e. Cryptographic Equipment. KY-65s, KY-57s, and KG-84s do not operate well in the cold. The expedient fix to this problem is to construct shelter. Place light bulbs/heat lamps in the shelter with the radio/cryptographic equipment to raise the ambient temperatures.

f. MRC Vehicles. The batteries of vehicle-mounted radio sets power the vehicle's starting system. Power for operating the radio is derived from the vehicle battery/generating system. The vehicles should be placed in as warm a place as possible. Both the radio and batteries should be well insulated on those sides that are against the cold metal of the vehicle body. To conserve battery life in the cold, operate the vehicle when operating the radio. Battery maintenance is important. Lead acid batteries should never be allowed to drop below two-third of full charge or charged below 15 degrees F. The specific gravity of batteries should be checked at least once a week using a view-type battery/antifreeze tester. Batteries with a specific gravity of less than 1,250 should be recharged and electrolyte added. When MRC vehicles remain shut down in temperatures below 0 degrees F for over 6 to 7 hours, operators must allow 10 to 15 minutes for the vehicle to warm and a constant idle to be established. Once the MRC vehicle is operating allow 5 to 10 minutes for the mobile radio to warm up. If the radio does not key out, the problem may be that the radio set is not warm enough. While the vehicle is shut down, start it every 1 to 2 hours and run for 5 to 10 minutes to prevent freezing of vehicle and radio components.

g. Electrical Tape. Electrical tape loses its grip in cold weather. Use low temperature tape, TL-600, NSN 5970-00-240-0620.

10. Environmental Effects. Environmental conditions will adversely affect performance of communications equipment.

a. Ice and Snow. Keep radios, radio remote sets, and cryptographic equipment off ice and snow. For remote radio antenna stations, use a 10-man tent or tent sheet shelter at the antenna station. Keep the radio equipment in the warm tent. If a tent cannot be used, fabricate insulated cold weather bags for radios and equipment.

b. Condensation. Condensation is a problem when temperatures fluctuate between freezing and thawing (25 to 40 degrees F). Radio equipment is susceptible to the same dangers from condensation as sweating; i.e., sweating and refreezing. Internal condensation caused by battery heat may take a long time to dry and may cause short circuits and damage. Additional radio equipment replacement items must be anticipated in cold weather operations because of condensation. Equipment must be replaced and dried out. This requirement will demand additional time and labor. Moisture from the breath will freeze onto handsets and quickly coat them in ice. The button or switch may also become ice-covered. Handsets should be waterproofed with plastic bags, ensure it does not interfere with the functioning of the handset. Remove frost from the equipment before bringing it into the tent.

c. Nighttime. Unless equipment is in temperatures above 10 to 20 °F, do not turn radios off at night if needed for operation in the morning.

d. Grounding: The use of high frequency radios requires establishing a ground. In cold weather, snow, ice, and frozen ground make it difficult. Each MRC vehicle must carry the following pioneer gear: a pick, a sledgehammer, Two ground stakes (4 to 6 feet long), and salt. Shovel the snow to the ice level; Use the pick to break through the ice and frozen ground; and pound the ground stake 2 to 3 feet deep, using the sledgehammer. Place a couple of handfuls of salt around the ground stake. The HF signal will be stronger and less interference will result if two grounds are used. Install the second ground stake within 2 feet of the first stake. Then connect the stakes with grounding wire. When communicating with the MRC vehicles, the vehicle should be in operation. For longer distances using HF communications (skywave), an AS-2259 antenna or expedient antenna may be used. A counterpoise may be required. Communications wire should be carried in the MRC vehicle for use as a counterpoise if needed. When necessary, an NVIS antenna can be constructed and carried in an MRC-138 or used on a PRC-104 to enhance HF sky wave communications in a mountainous environment. In areas like Southern Norway or Northern Germany where trees are plentiful, a nail may be driven into a tree and the equipment will be grounded by using the tree's root system.

e. Atmospheric Interference

(1). Ionosphere Disturbances and Auroral Effect. Arctic regions are often areas that experience ionospheric disturbances (sunspots and solar flares) and auroral effect (aurora borealis or northern lights) interference. This interference causes static during all types of radio communication, particularly HF radio signals. The least effected signals are ground wave signals. For example, MRC-138 using the 32-foot whip antenna, PRC-104 using 10-foot whip antenna, and VHF systems, which are line of sight signals.

(2). Radio Wave Propagation Charts. The Electromagnetic Compatibility Center, Annapolis, Maryland publishes radio wave propagation charts recommending minimum and maximum frequencies for the best communications during various hours in a designated day. Frequency planning is important and spare frequencies should be used for designated time of the day and night for HF communications. Tactical air request, naval gunfire, and higher headquarters HF circuit frequencies must be planned.

(3). Precipitation Static. Precipitation static is interference caused by snow, ice, and rain particles driven against metal objects and antennas. The result is a static noise that will interfere with HF, VHF, and UHF signals. Relocating radio equipment away from vehicles, buildings, and trees can reduce precipitation static. Precipitation static is associated with metal antennae, high power radios, and sensitive receivers exposed to rain or wet snow. Covering the antenna with polystyrene masking tape will reduce this effect, but it will only be effective if there are no other adjacent metal surfaces (sides of a vehicle, etc.) against which the discharge can take place. This is not apt to be a problem at battalion level and below.

11. Rebroadcast Stations. In mountainous terrain, rebroadcast stations with relay or rebroadcast teams will be needed. The normal T/O and T/E will not provide enough communications assets or personnel to establish all the necessary rebroadcast stations. Additional radio equipment and radio operators will be needed to set up and operate the relay or rebroadcast sites. Requirements must be identified to higher headquarters early in the planning process. Multiple rebroadcast stations may be required to bounce radio waves around the mountains. Options including floating AAV C-7s, remotely piloted vehicles, and

helicopter borne platforms may be considered. When using AAVs, the ice makes for a great takeoff and allows VHF waves to expand before hitting the valley.

a. Camouflage and Concealment. Picking suitable relay sites is challenging and probably will deviate from the normal practice of finding the tallest mountain in the area. In the cold weather operating area the tree line will occur will below the top of the mountain which leaves the rebroadcast station exposed and without the ability to be adequately camouflaged.
b. Equipment and Personnel. Equipment to provide for two VHF nets and one HF net will be required. Because of distances covered, RC-292 antennas will normally be required. The sites will normally be manned by a four-man team, which will require adequate equipment to survive.
c. Insertion / Extraction. Because of the heavy equipment load required to operate a rebroadcast site, the team operating the site will require additional support in order to haul necessary equipment during insertion and extraction.
d. Training Rebroadcast Teams. Personnel in four-man rebroadcast teams must be capable and self-sufficient. At least one man must have exceptional knowledge and experience. All should be snow mobile trained and be able to move equipment using sleds. Teams should be trained in helicopter insertion/ extraction, and in the use of expedient shelters and cold weather camouflage/ concealment.
e. Recommend Augmentation for Rebroadcast Teams. Recommended T/O for augmenting the infantry regiment for retransmissions is 1 officer and 16 enlisted with a minimum of 3 relay teams. Appropriate additions of equipment will also be required.
f. Continuous Waves. Continuous Wave (CW) keying devices may be used to send Morse code on HF circuits when the quality of voice becomes unreadable. Radio operators must be trained in CW. A brevity code can be designed to switch to CW.

12. Operator Maintenance. Because the polar regions are subject to disturbances, which affect radio reception, it is important to get the very best performance from radio sets. Operators must be intimately familiar with their sets and should keep radio equipment *clean, dry, and where possible, warm.* They should handle the set and its ancillary equipment carefully, knowing that most materials become fragile at low temperatures. Maintenance of the set and batteries should be regular and meticulous. Reports of any defects should be made as soon as they are noticed. The main points which should be covered are:
    - Always keep plugs and jacks clean.
    - Antenna connections must be tight.
    - Keep insulators dry and clean.
    - Always remove snow and ice.
    - Power connections must be tight.
    - Motors and fans should turn freely.
    - Knobs and controls should operate easily.
    - Keep dry batteries fresh, warm, and spares on hand.
    - Install breath shields on all handsets.

- Coat cables and wires with silicone.

13. Wire (Telephone) Communications. Wire (or telephone) is the most preferred type of communications in the cold because there are fewer problems with actual communications. However, installing and maintaining the network is time consuming, manpower intensive, and creates several problems.
    a. Battery Power. The BA-3030 (cold weather) dry cell battery must be used in field telephones. These batteries must be protected from the cold. Do not place field telephones on the ground or in the snow.
    b. Material Failure. Like radio equipment, telephone equipment becomes brittle and is very susceptible to breakage when temperatures are below ten (10) degrees Fahrenheit. Common problems are:
        - Field telephone handset cables must have a thin coat of silicone lubricant.
        - TA-312 field telephones provide the best wire communication. A microphone moisture cover must be installed in the telephone.
        - TA-1 sound-powered telephones have a carbon element microphone, which freezes and needs to be kept warm and dry to operate.
    c. Wire Laying Techniques. Laying the wire may be done in a variety of ways. The standard methods may be employed such as by wheeled vehicles and helicopters. In a snow-covered environment, with Marines on skis or snowshoes, over snow vehicles may be necessary. The wire will normally be laid above the snow whenever possible to prevent losing sight of the wire and for ease of laying and retrieving. When laying wire across roads, the recommended procedure is to elevate the wire above the road. If the wire must be placed on the road, special care must be taken to ensure the wire is buried well below the running surface. If there is an absence of snow and the frozen ground prevents burying the wire, plan an alternate means of covering the wire (sandbags or lumber).
    d. Cable Installation. If cable must be installed, mark each of the 26 pairs of cable connectors in the event snowfall covers them. This technique will aid in troubleshooting and cable recovery.
    e. Host Nation Support. The use of host nation cable system, when possible, will conserve tactical wire and cable. Host nation base cable systems are usually underground and provide better quality lines with less impact from the weather.
    f. Personnel. Because of the problems associated with radio communications in cold weather and mountain operations, wiremen will find their tasks exceptionally tough. All wiremen should be snow mobile and knowledgeable in using expeditious shelters and survival techniques. Additional personnel will be needed and must be planned for.

14. Visual Communications. Visual communications is an accepted method in most situations. However, in a cold weather environment, it will be ineffective during the periods of reduced visibility caused by long winter nights, fog, and whiteouts. Visual signals should be prearranged and in the operation order. The standard air panel markers contain 1 set of white and black markers with 13 markers per set.

White markers are useless in snow-covered terrain. Replace them with another color. Semaphore flags (red and white) must have the same consideration. Pyrotechnics must be prevented from sinking into snow. Attach them to some sort of platform.

15. Preventive Maintenance. Preventive maintenance is essential for proper operation of communications equipment. The communications contact teams and headquarters groups will need to have additional maintenance personnel attached. They will have sufficient amounts of pre-expended bin items and supplies, such as handsets, coaxial cable, connectors, whip and base antennas, etc.
16. Safety Precautions for Operating Equipment.
    a. Temperature. If temperatures are below +10 F, do not touch metal parts with bare hands.
    b. Wind. Construct antennae to be secure against blowing wind.
17. Equipment Load.
    a. Planning. During cold weather operations, communications personnel will need to carry communications equipment in addition to their required personal equipment. **(PEB Pending)**
    b. Methods of Reducing Load.
       (1). Use the logistic trains to provide resupply of batteries, wire, preventive maintenance material, waterproofing material, and maintenance support for interchanging equipment that is inoperable.
       (2). Spread load the communications equipment and cold weather equipment between Marines. (REALIZE THAT MARINES ARE ALREADY HEAVILY LOADED.)
18. Communication Variant Vehicles. The AAV C-7 and LAV may provide additional radio equipment not identified in the infantry regiment T/E. They also provide heated areas and an additional mobility capability.
    a. AAV C-7. The C-7 will provide a capability to move in the unfrozen waters of fjords and streams. If snow is considerable, the AAV will be roadbound. However, these vehicles are relatively mobile and displace rapidly.
    b. LAV. The LAV, equipped with chains, is fast and mobile both on and off the roads in moderate snow depths. Self-recovery capabilities of LAVs make them ideal for quick displacement. In deep snow and on icy roads, they will be roadbound.
    c. BV-202 / BV-206 C2 Variant. These vehicles are propositioned in Norway and provide exceptional CP/CP displacement capabilities. Their off-road capability is unsurpassed. The BV-202s, which are dedicated to the United States Marine Corps, are pre-positioned in Norway. They come with drivers/signalmen who can be used to augment USMC needs, especially in maintaining communications with allied units. Radios used in the BV-206 C2 variant will need to be provided from the unit's T/E.

5003. Supply Considerations.

A. Special Supplies. Supply requirements include the procurement and distribution of many additional specialized sub-classes of supply, which are needed for cold weather operations. This includes items found in the Contingency Training Equipment Pool (C/TEP) and/or in war stock. These requirements must be determined for each exercise/operation according to terrain, weather, time of year, and nature of the operation. All classes of supply are affected by the cold, some to a much greater extent than others. Through anticipation, redundancy, imagination and innovative solutions, the logistics officer can meet the MAGTF's needs.

B. Unit Distribution. A good distribution system reduces the size and number of supply dumps, minimizes surges, and provides rapid, flexible support to the landing force. The supply process must be responsive to the needs of the supported unit, especially in a cold weather environment. Centralization may be inefficient and unresponsive, and may reduce survivability. In cold weather conditions, unit distribution is preferred. This allows the commander the flexibility to decide which transportation system will be employed; i.e. ground or air. The most responsive unit that delivers supplies to subordinate elements is the mobile combat service support detachment (MCSSD). MCSSDs are the link between the forward and subordinate elements of the tactical units and the supporting CSSE. They provide responsive support at the forward-most point. In cold weather environments, MCSSDs can become the deciding factor if weather conditions deteriorate and transportation becomes limited and/or nonexistent. Supplies that are preloaded on the MCSSD's transportation assets could determine whether missions are accomplished or aborted.

C. Repair and Replenishment Points. Planners must develop a flexible system that can react to changing environmental factors and provide the essential support for the landing force. Organizations must rely heavily on organic capabilities, applying the train concept. In coordination with the operations officer, the logistics officer must plan the locations of unit trains. Preplanned repair and replenishment points (RRPS) are established primarily for classes I, III, and V. Company/battery level unit trains obtain their re-supply from the CSSD combat trains at these locations, normally during the hours of darkness, with each unit allocated a specific time period. Unit trains use organic off-road mobility assets (AAV, BV-202/206, and LAV) to maintain accessibility to the supporting CSSD or the battalion trains that are often road bound.

D. Location of Supply Dumps. Supply dumps for critical classes of supply (I, III, and V) and selected items in classes VII, VIII, and IX must be located closer to the front than normal to be responsive. This was illustrated in Korea during October -December 1950 by the 8th Army and the 10th Corps (1st MarDiv). Before the Chinese entered the war, the Marines were criticized for being slow in their push north. Marine General Oliver P. Smith was careful not to move his forward line of troops beyond his ability to provide for their support. This eventually proved critical and provided for the Marines as they fought their way successfully out of Chosin using these pre-staged supplies. Conversely, the 8th Army was extended well beyond its supply dumps when the Chinese attacked. Its retrograde was poorly supported and unorganized and was referred to by Brigadier General S.L.A. Marshall as *The Great Bug Out.*

E. Classes of Supply

1. Class I, Subsistence.

    a. A basic fact of cold weather operations is that Marines must eat more than usual to function. The greater part of what a Marine eats and drinks will help maintain their body heat while a small proportion produces energy for physical work. The average Marine's caloric need will increase from an average of 3,900 to 4,500 - 6,000 calories per day. Additional calories are provided through four meals-ready-to-eat (MRES) per day when field duty is in excess of 3 days or by eating food supplements. The cold weather rations (RCW) (also called the arctic ration) is designed specifically to meet cold weather feeding requirements. The RCW provides 4,500 calories per day and contains 14 components that do not freeze. It consists of two packages including entrees, snacks, and numerous hot drinks. The tray pack rations (T-ration) contain entree, vegetable, starch, and dessert meal components. T-rations are precooked, thermostabilized bulk food items in sealed half-size steam table trays. They can be heated just before eating.

    b. Water re-supply is a constant problem when temperatures are below freezing. Special provisions are required for potable water production, insulated storage, and distribution. The re-supply effort may require using heated compartments of over snow vehicles to keep water from freezing. Close coordination is necessary with the receiving units so that the water is distributed before it freezes. If water is to be provided by melting snow with fuel-fired stoves, more fuel must also be provided. The M-146 water trailer heaters have sometimes been unreliable and if used, should be closely monitored. Reverse osmosis water purification unit (ROWPU) for water production has inherent problems in the cold; i.e., reduced production and freezing of filters. Use a heated tent to cover the ROWPU in extreme cold. The Naval Construction Force (NCF) well drilling assets may be needed to provide water for sustained operations. Plastic 5-gallon water cans that are currently in use will break if filled at extreme temperature. When transporting water, keep them only partially full and turned upside down.

2. Class II, Clothing, Individual Equipment, and Tents

    a. Units should plan on sufficient extra individual cold weather clothing items to replace lost or damaged clothing. Units should be able to react to both dry cold and wet cold conditions.

    b. The need for tent heaters and stoves in storage areas must be emphasized. Maintenance type tents will have to be embarked along with Herman-Nelson heaters. Heated tents will also be necessary for the storage of some classes of supply.

    c. Unique requirements must be anticipated at the unit level; e.g., avalanche cord, ski wax, candles, axes, shovels, matches, sunscreen, tire chains, and winterization kits. The primary sources for this equipment are the C/TEP or war stocks. Marines will have to carry their own fuel for squad stoves. This will require special V2-liter bottles. These bottles must be filled using either small funnels or special fuel cans with pour spouts. Sunglasses, special fuel containers, cold weather clothing items, and tentage may be available from

C/TEP. (See FMFM *7-23, Small-Unit Leader's Guide to Cold Weather Operations,* chapter 2, for more information on clothing and personal equipment.)

3. Class III, Petroleum, Oil, and Lubricants (POL). Vehicles will have to be operated continuously when weather is at -25F. A dramatic increase in POL/fuel requirements must be anticipated. Fuel and lubricants will need to correspond with changes in temperature.
    a. Travel over snow or tundra can increase fuel consumption by over 25 percent. Diesel fuel will need additives to prevent freezing and gelling of fuel. This is a bulk fuel responsibility. A substitute is JP-5. Correctly mixed fuel will normally be available in the cold weather operating area but not at the embarkation point. Standard diesel fuels will not work in sub-zero temperatures. Marine Corps vehicles have multi-fuel capability. If fuel with additives is not available at point of embarkation, use JP-5.
    b. Multi-viscosity oil (15W-40) is recommended for most rolling stock in the cold. Use of 15W-40 will preclude a need for frequent oil changes in an environment with wide, rapid temperature changes. Vehicles should be changed to multi-viscosity oil before embarkation. Only in sustained extreme cold conditions will 10W oil be required. Units should embark a block of 10W in sustained cold.
    c. Special fuels may be needed if using host nation equipment. For example, the BV-202 requires gasoline of 95% octane. The standard DOD 80 octane gasoline will foul its carburetor jets. Generally this fuel is available through host nation support.
    d. Higher levels of class II (containers) to carry or store the fuel should not be overlooked. Up to double the normal number of fuel cans may be needed if carrying fuel to the vehicle instead of bringing the vehicle to the refueling points. Even at extremely low temperatures, fuel can be delivered in 500-gallon fuel bladders (drum, fabric, collapsible liquid fuel, 500 gallon).
    e. References for fuel problems include:
        - Military Handbook, Mobility Fuel Users Handbook 114.
        - FM 10-69, Petroleum and Fuel Operations
        - TI 10340-15/lA, Fuel Compatibility for Engines, Motor Transportation and Ordnance Tracked Vehicles.
        - TM 3835-15/1, Fuel Testing Standards.
4. Class IV, Construction Materials. The lack of ground systems such as roads, airfields, ports, and railways places greater demands on providing and moving building supplies. To develop and maintain MSRs over both land and water, there will be an initial demand to construct both storage and living spaces. Supplies needed to maintain MSRs, roads, and bridges must be determined for each operation according to terrain, weather, and the nature of the operation. The cold weather environment requires more engineering supplies than normal because of the uniqueness of the area. Staging areas in prospective destinations are nonexistent/inadequate. Offloading must be selective and highly organized. Supplies must be combat-loaded.

a. Great quantities of gravel will be needed for roads and airfields to cover vegetation of permafrost areas. Bulk gravel or sand must be covered with plastic sheeting or canvas so they can be worked below freezing temperatures. Salt also serves to prevent freezing, however it is highly corrosive.
b. Dunnage needs to be embarked for it is not available in remote arctic areas. Indigenous shipping and receiving sites may be a source of dunnage.
c. Construction supplies placed directly on the ground may freeze in place. Expendable material such as canvas, cardboard, or dunnage should be placed between items or pallets to be stored on the ground.

5. Class V, Ammunition. There will be a greater need for ammunition. The CSSE can anticipate handling, storing, and moving greater volumes of ammunition and demolitions. Special storage for ammunition will not be required but it should be stored in original containers. Preparing ammunition dumps will be more difficult because of freezing and mud conditions.
6. Class VI, Personal Demand Items. Morale in the cold can be significantly enhanced by paying attention to the health, comfort, personal demand, and PX-type items found in the class VI block. Demand for this class of supply remains constant in cold weather operations as in any other operation. Care must be taken to avoid stocking items that will freeze.
7. Class VII, Major End Items. During cold weather operations, an emphasis must be placed on preventive and corrective maintenance. First and second echelon maintenance are unit responsibilities and will become more important. It may not be feasible to evacuate equipment to the rear because of the limited and/or narrow road systems. It may be advisable to assign a CSS liaison to advance with the forward elements and to send a forward mobile CSSD in trace of the GCE. Marines in these billets need to be highly skilled in supply and cold weather operations. They must constantly monitor events to anticipate a unit's re-supply requirements. Replacement of major end items when possible is preferred. However, movement of deadline equipment over snow and ice will further stretch the maintenance capabilities of a unit. This may clog the limited MSR. An increased demand for power generators, Herman-Nelson heaters, and rough terrain loaders with a snow removal capability should be anticipated
8. Class VIII, Medical Supplies. High consumption rates of medical supplies must be anticipated to include an increased need for such items as chapstick, cough syrup, and decongestants. Heated storage areas (warming tents, vehicles, and containers) are required for storing liquid medications and whole blood. Solid medications and freeze-dried material instead of liquids can be used when building the AMAL to minimize freezing, storing, and handling problems. Perishable materials must be packaged and marked for special handling. Procedures must be established and followed for special handling of the AMAL from embarkation through its final destination. (**Supplemental AMAL pending**)
9. Class IX, Repair Parts. The cold will significantly affect the need for spare parts. Metals and synthetics become brittle; batteries fail. Tools break and get lost in the snow, especially when operating in long periods of sustained, extreme cold (below -25 °F). The need of spare fuel can gaskets must be anticipated. Complete equipment LTIs must be conducted before embarkation. Greater

attention must be given to requesting class IX repair parts before deploying. Especially important are high demand items; e.g. starters and alternators. Equipment will be under unusual stress. Tires wear from chains; battery cranking power is reduced; and metals break. Units should anticipate the requirement for high parts usage and have them prepositioned as far forward as possible. Dry cell batteries should be given special consideration. They should be new and replacement factors closely examined. The specific gravity of wet cell batteries must be constantly monitored. See paragraph 5002 for a discussion on communication batteries.

10. Class X, Material to Support Nonmilitary Programs. The cold weather affects all personnel, including enemy prisoners of war, indigenous inhabitants, and refugees. The same protective measures and commodities necessary to protect friendly forces are needed. Considerable effort and expenditure will be necessary to provide for native inhabitants and refugees taken into custody. The numbers of nonmilitary personnel should be kept to a minimum whenever possible. All classes of supplies to care for noncombatants should be anticipated to meet this requirement in cold weather operations.

F. Storage and Movement: Special storage and movement concerns must be considered when bringing large quantities of supplies into a cold weather operating area. These considerations include the following:

1. Location: Storage areas should be located as close to all-weather roads as possible.
2. Distance: The distance to supported units must be as short as possible since time and distance are key factors and MSRs limited.
3. Freeze/Thaw. Site selection must consider the effects of periodic or seasonal thaws on the accessibility and condition of stored supplies. Storage boxes that provide dry storage capabilities for items needing protection from the weather must be procured.
4. Mobile Loading: Classes I, III, V, and IX should be mobile-loaded or on pallets (bulk packaged) for easy accessibility.
5. Temperature: Heated areas may be required for several classes of supply (class I water, classes VII and VIII). Medical supplies, special fuses, batteries, and other items subject to damage by freezing must be stocked in heated shelters. Refrigerator boxes may be used when turned off to keep supplies warm provided a heat source can be improvised.

G. Identification: Supplies stored in open areas that may be subjected to drifting snow should be marked with poles and small flags. Particular attention must be given to marking medical supplies.

H. Staging and Marshaling Areas: Staging and marshaling areas are extremely limited in maritime cold weather operating areas. Mountains begin below sea level and leave little flat/rolling terrain at the waters edge. Beaches are extremely limited and may be clogged with ice or windblown snow. Development of staging areas will be difficult because of frozen ground, frost, permafrost, poor drainage patterns, and previous development of almost all of existing terrain for civil use. Previously developed civilian facilities may be considered for development as staging and marshaling areas.

I. Host Nation Support (HNS). HNS in many cases can significantly reduce embarkation and lift requirements for virtually every class of supply. Special consideration should be given to food supplements (perishables), fuel (bulk and package), lumber, medical supplies, batteries, and some repair parts. Every item procured through HNS is one item less that will require shipping, handling, and storing.

5004. Motor Transport Considerations.

A. Estimate of Support: The logistics officer must fully understand his transportation capabilities when developing his estimate of support. Mission accomplishment may be subject to the types of transportation assets, their availability, and the influence of the weather on their mobility. Convoy operations and re-supply methods must be creative and unpredictable because of the nature of the terrain and the fact that trafficable roads are few in a mountainous or arctic environment. Furthermore, these roads can be easily targeted by enemy aircraft or used as ambush sites. In most cases, transportation will be provided by motor transport assets and, weather and altitude permitting, helicopters. Redundant methods of transportation must always be planned. Fjords, rivers, streams, and frozen rivers/streams that may be obstacles in the summer months can be considered MSR's during the winter. Use host nation resources when available. The Small Unit Support Vehicle (SUSV), also known as the BV 206, may augment organic assets, motor transportation, and the Logistics Vehicles System (LVS), a marginal terrain tracked vehicle that has excellent mobility in snow covered terrain. The LVS is good for moderate terrain, but has a limited off-road capability in snow and mountainous terrain. AAV's can be used in fjords, on plowed roads, and to a limited extend, cross-country.

B. Planning for Snow Removal. Snow can quickly become an obstacle for any vehicle and must be anticipated in any cold weather or mountainous environment at any time of year. Efficient and timely snow removal must be planned for well in advance. Snow removal may be a host nation responsibility but in some theaters, this task must be planned requesting Marine Corps' engineer assets, U.S. Army engineers, or U.S. Navy mobile construction forces.

C. Safety and Survival:

1. Fuel Handling. In arctic conditions fuel spilled on flesh can cause instant frostbite. Uniforms soaked with oil or fuel conduct heat away from the body, much like water, and are a fire hazard so they must be cleaned immediately. All of the layers in the ECWWS clothing system must be kept as clean as possible to allow the garments to work effectively and POL's will reduce their ability to retain heat and transfer water away from the body.
2. Survival Training: Cold weather survival training is necessary for all personnel. Whenever a vehicle is dispatched, the crew should be provided with radios and an emergency radio frequency. Each vehicle must have survival equipment on board for all personnel.

D. Camouflage and Concealment: Camouflaging equipment can be difficult in arctic conditions. Whitewash and latex paints are not authorized because they decrease effects of the Chemical Agent Resistant Coating (CARC). When camouflaging remember:

- Use the proper camouflage pattern
- Avoid brilliant whites.
- Don't paint canvas. The paint will make the material brittle, does not wash off, and will ruin what little water repellency the cotton duck possesses.
- Only use CARC paint or paint authorized in the TM.

- Break up outlines and cover all reflective surfaces.
- Dig in vehicles whenever possible.
- Camouflage nets that cover the entire vehicle should be set up to break up the vehicle's outline as much as possible. Pay attention to noise, infrared, and visual means of detection. Remember that engine exhaust gives off smoke, and in extreme cold can turn to ice fog.

E. Transportation Support:

1. Air Delivery Operations. Refer to Chapter 4, Air Operations.
2. Port Operations. During cold weather operations, port facilities become hazardous areas because of snow and ice build up on already slippery ramps and gangplanks. Since tire chains are not permitted on ship, salt or cinder may have to be used for traction. Strict traffic control is necessary because of the hazards of moving vehicles in tight spaces on icy surfaces and to minimize confusion. Ground guides and good communication procedures will help control vehicle traffic. Ensure that personnel are rotated frequently and given time to get out of the cold weather. If the ship is not willing to allow Marines to use their facilities, warming shelters should be used. Maintenance contact teams must be in place on-board ship and at the port staging areas to assist in the startup of equipment. Medical personnel should also be on-hand.
3. Material Handling Equipment (MHE). The same considerations for motor transport equipment must be given to MHE because this equipment is affected in the same way by the cold and in some cases, many MHE assets are more susceptible to maintenance problems because of their reliance on hydraulics. Hydraulic systems should be routinely drained, cleaned of all contamination, and refilled with proper fluids to ensure dependability. Some MHE assets may be used for snow clearing. They should be sheltered from the snow as much as possible when not in use or kept running continuously to avoid cold related problems.
4. Ferries and Landing Craft. Ferries and landing craft should be used when possible. The same considerations apply for ferries and landing craft as found in Port Operations.
5. Host Nation Support. When available, host nation support is a force multiplier that can be used to move equipment and personnel without having to use tactical equipment. Host nation support is limited only to the imagination and can save Marines time and effort and wear and tear on vehicles if it can be procured.

F. Helicopter Support. Helicopters provide a flexible and responsive means of transportation when not grounded by inclement weather or cold weather related maintenance problems. Therefore, motor transportation assets must be on stand-by to accomplish any requirements that aviation may not be able to meet.

1. Forward Operating Bases (FOB's) and Forward Arming and Refueling Point (FARP) Operations. FOB's and FARP's may be needed to develop more responsive solutions to transportation or re-supply situations. See Chapter 4.
2. External Operations. External operations are particularly difficult in cold weather. Wind-chill, visibility, and static electricity must be considered (static electrical discharge from helicopters can be fatal to ground support personnel, see

FMFM 7-22 and FMFRP 5-31, Multi-service Helicopter External Air Transport Procedures and the aviation section of this publication).

G. Motor Transportation. Doctrine, techniques, and procedures for motor transport operations are contained in FMFM 4-9, Motor Transport. Another good reference is FM 55-30, Army Motor Transport Units and Operations. The concepts found in this MCWP do not differ dramatically from those publications.

1. Personnel Considerations:
   - Always use the buddy system.
   - Operators should be aware of the possibility of ice build up on the soles of boots and brake/accelerator pedals.
   - Personnel must not be allowed to sleep under or in vehicles.
   - Operators should be aware of the possibility of carbon monoxide build up in the cab of a vehicle that is surround by deep snow.
   - Maintenance facilities should be as protected from the elements as possible.
   - Marines must wear gloves to prevent hands from freezing to metal and when handling any POL's.
   - Unit leaders must constantly check for the signs and symptoms of cold weather related injuries.
   - Warming tents should be used to rotate Marines out of the elements.
2. Cross Country Movement. The problems of cross-country movement are significant for conventional motor transportation assets. Unless snow covered areas are cleared of snow and ice, most vehicles will not be able to move cross-country even with chains. A majority of vehicle traffic will be restricted to improved surface roads that have been cleared of snow and ice by engineer or host nation support.
3. Vehicle Preparation: All equipment must be prepared for cold weather operations prior to arriving in theater. All leaders should consider the following:
   a. Cold weather kits are necessary for every vehicle. They should include at a minimum:
      - Tire chains for all wheels
      - Tire chain repair kit
      - De-icer, non-freeze windshield wiper fluid, scrapers
      - Tow bars or straps
      - Extra chock blocks
      - Plastic or extra canvas to cover windshields to reduce buildup of ice or frost.

   b. If possible, extra operators and mechanics should be taken into a cold weather environment to minimize down time because of reduced efficiency in a cold weather environment.

   c. Extra tents and heaters should be embarked for motor transportation personnel.

   d. Repair parts blocks should be increased, especially for parts that are susceptible to the cold such as starters, generators, alternators, glow plugs, etc.

H. Operations: Vehicle operation is more difficult and the reliability decreases in colder temperatures. Special considerations must be given to keeping engines warmed up and out of the weather, if possible, or started regularly to avoid some problems.

1. Chains: Every vehicle should have a complete, serviceable set of chains for all wheels. Because chains break frequently, chain repair kits should also be carried. All motor transport personnel operators should know how to properly put chains on any vehicle. Additional chains are required for trailers and artillery pieces.
2. Ditches and Shoulders: All operators must be aware of the dangers of drainage ditches and soft shoulders present in most areas where heavy snowfall is expected. These areas are used to help drainage and can become easy traps for vehicles that stray too close to the side of the road. Although snow stakes are good indications of where the road is, they can often lead one too close to the side of the road and into a ditch.
3. Parking: Whenever parking in cold, wet conditions use some kind of dunnage, tree branches, wood, MRE boxes, etc., to park on to prevent tires from freezing to ground. If tires become frozen to the deck, antifreeze or fuel may have to be used to free them. Always park where a vehicle can be easily towed and slave receptacles are within easy reach.
4. Emergency Brakes: Use chock blocks instead of emergency brakes when parking. Emergency brakes can easily freeze when set. Always drain air tanks when parking a vehicle to prevent condensation from freezing.
5. Starting and Preheating Engines: Establish times at regular intervals to start vehicles, at least every three to four hours. Vehicles should be run for at least ten minutes or until normal operating temperature is reached with the air cocks open to prevent them from freezing. Once the vehicle is shut down, shut the valves.
6. SL-3 Equipment: All SL-3 gear should be in good working order and augmented with the equipment listed in paragraph 3 above.
7. HAZMAT and POL's: More HAZMAT and POL's will be used in a cold weather environment that normal operating conditions. Fuel consumption will increase 25 – 50% and more anti-freeze is required if radiators have to be drained or to replace lost fluid.
8. Special Considerations for the M900 Series Vehicles: The M-900 should run approximately 15 minutes with the air cock open to rid the system of moisture and to ensure adequate circulation of alcohol. Heaters should not be run while the M-900 is idling because the air-cooled electric motor could burn out. When possible, park the radiator away from the wind to reduce heat loss. Do not exceed 15 mph when in low transfer and tires should be inflated, or deflated, to cross-country pressure.

I. Maintenance Procedures.

1. Operators: As always, pre, during, and post-operations preventative maintenance is the operators' responsibility and will take longer in a cold weather environment.
2. Maintenance: Scheduled maintenance takes on a greater importance in a cold weather environment than under normal operating conditions. All scheduled maintenance must be completed to reduce corrective maintenance. Adequate shelter must be constructed to provide mechanics an area to work that will not lead to cold weather injuries.

J. Driver Training. Because most motor transport Marines will not have cold weather operating experience, training is needed for all personnel. At a minimum training should consist of:

   1. Survival Training.
   2. Basic principles for driving on ice and snow, practical application if possible. Some training considerations:

- Work from easy, snow covered terrain to difficult driving conditions.
- Remind Marines that even with 4-wheel drive, stopping distance may not be decreased. Braking distance is generally doubled and increases with the amount of weight being carried.
- Ground guides can be used when navigating sharp bends and turns and especially when pulling to the side of the road.
- Vehicle recovery, including self-recovery.
- Road rules, signs, and regulations.
- Tire chain installation.
- Cold weather vehicle operation and maintenance (1st and 2nd echelon).

K. Road Master Responsibilities. The road master must supervise the safe operations of all vehicles and work closely with PMO to identify danger areas, keep units informed of road conditions, monitor speed limits, etc.

L. Off-road Transport. In some winter theaters, the Small Unit Support Vehicle (SUSV), also called the BV 206, may be available to move Marines and supplies. These vehicles are marginal terrain vehicles that can operate in snow up to 3 1/2 – 4 feet and over any road with packed snow.

5005. Marshalling Considerations.

Marshalling is the preparation of military personnel or equipment prior to embarkation or their subsequent organization and distribution following debarkation at a port or airhead in a theater of operation. This section deals with the latter (debarkation) in a cold weather theater of operations.

A. Space: Consideration must be given to two things regarding space.
    1. Cold weather environments are sometimes found in mountainous terrain, which do not lend themselves to wide open areas required for the debarkation and organization of an offloading MAGTF. Theater logisticians must have enough advanced planning time to figure out where the massive amounts of equipment and bulk stores will go during offload.
    2. Logistics Movement Control Center (LMCC) must consider how frozen precipitation will affect their staging areas. Certainly engineering equipment (primarily blades) will be needed to clear parking areas but piled snow/ice takes already limited space at the airhead or port.
B. Throughput: The ability of LMCC to effectively perform the offload and distribution of the equipment and supplies to support the MAGTF will be limited by two things, space at the airhead/port and limited road infrastructure from the airhead/port.
    1. Time: Because of limitations, the process of marshalling can be approximately doubled.
    2. Multiple debarkation points: If feasible, operating multiple logistics hubs instead of one huge ALOC will ease the pressure on limited road systems. Additionally, it decreases the footprint of the logisticians and makes the rear area less identifiable.
C. Summary: Time and advanced reconnaissance are critical to the task of marshalling the offload of the MAGTF. The marshalling task will require the logistician to solve two overriding questions: Where will the equipment go? How quickly can we move it through the control point?

5006. Maintenance Considerations.

A. Time and Distance: Maintenance will take up to five times longer. Time spent on maintenance tasks will be proportionate to be the availability/nonavailability of adequate facilities. Under the most adverse environmental conditions, if facilities are not available, certain tasks may have to be postponed until more favorable conditions exist. In the cold, this can be weeks. When equipment has become cold-soaked, extensive time and facilities will be needed to warm equipment even before repairs can be started. Distance between units, from the breakdown point to the repair point, or to an available heated maintenance facility will often be great and will adversely affect performing maintenance. Maintenance facilities will likely be spread over a large area compounding communications and control problems. Facilities must be heated and wind resistant.

B. Combat Operations.

1. Offensive Operations: Supporting units will be located in rear areas and will be engaged primarily in rehabilitation of damaged equipment. Recovery and on site repair by forward contact teams will be exploited to the maximum extent possible. Recovery capabilities will be adversely affected by heavy snow, extensive muskeg areas, unpredictable weather, and a limited road network.
2. Defensive Operations: Ordnance support is forward with supported units on maintenance assistance missions.

C. Preventive Maintenance: The need for preventive maintenance is increased in the cold. Units must perform preventive maintenance before their equipment is embarked. During embarkation, ship's platoon operators and mechanics should start vehicles during movement to the objective area on a scheduled basis. In the cold weather operating area, there is a constant need to exercise equipment to prevent it from becoming cold soaked. Never shut down all assets at the same time. A rotating system needs to be established where a vehicle is used to start another before it is shut down. Increase the frequency of operational checks. Conduct pre-inspections, post operating inspections, and daily operational checks at every echelon.

D. Preparing Vehicles for Cold Weather Operations. The extreme cold is tough on equipment and component parts. Most vehicles are designed to operate in temperate climates and must undergo winterization to function properly in the cold. TMs and LOs must be closely followed. Back planning time is necessary to prepare vehicles for cold weather operations. This is a first echelon responsibility. Tires, batteries, cooling systems, and lubricating/fluid systems must receive special attention. Equipment being transported by rail, roll-on/roll-off ships, or by MAC aircraft must be prepared for winter operations before embarkation. Once winterized, these vehicles may continue to operate year-round in northern operating areas using cold weather POL.

1. Tires. If possible, units should modify 6x6 trucks to run on super singles with chains or studs rather than duals. This provides added traction to the vehicle on ice and in snow. All tire stems should be capped to prevent moisture buildup.
2. Batteries: Batteries at one-third or less charge may freeze and should be replaced before embarkation. Specific gravity must be constantly monitored. Batteries must be kept warm and fully charged. The 12-volt batteries have a higher than

normal replacement factor in cold climates. They lose cranking power rapidly as temperatures drop below freezing. Battery blanket warmers are available in the supply system but they must be requested. When temperatures fall below -25 °F and the vehicle will not be operated for more than 8 hours, remove batteries and store inside a warm enclosure off the ground.

3. Heaters: Units should insure that all vehicles have effective personnel compartment heaters installed. Herman Nelson-type heaters are prescribed for maintenance tents and other large utility structures.
4. Winches: During extremely cold weather, metal becomes brittle and breaks easily. Reduce the designated capacity to a level that will not damage the winch or the vehicle.
5. Fluids and Lubricants: Always consult the appropriate TM or LO. Remember to use antifreeze in coolant systems. Different grades of hydraulic fluids are also necessary. If vehicles come into theater with temperate grade automatic transmission fluids and becomes cold-soaked, seals will likely blow when the transmission is engaged.
6. Fuels: Failure to use proper fuel in vehicles at -20 °F or colder will result in waxing of the fuel in tanks, carburetors, and fuel lines. The vehicle must be placed in a heated area until thawed and all fuel and fuel lines purged. This procedure is time consuming and space demanding. **The option exists to use JP-5.**

E. Personnel: The wind and cold will drain personnel. When task organizing for operations in the cold, the commander must realize that productivity of the individual will diminish because of the cold's drain on the body. Additional maintenance personnel must be requested. There will be a constant leadership challenge to prevent cold weather injuries when conducting maintenance in the cold.

F. Repair Parts. The cold temperatures will have an adverse effect on the durability of parts and tools. Test equipment will often be unreliable. Equipment and tools will become brittle. Differences in coefficients of expansion and contraction can cause failures and false readings. There should be an increased reliance on secondary repairable, component repairs, and selective component exchanges. More fuel filters will be needed in combating fuel line icing. Repair parts should be warehoused or stored near repair facilities if possible.

G. Evacuation versus Reclamation: A cold-soaked/inoperable vehicle can seriously limit operational capabilities and present a challenging situation to maintenance. Alternatives include attempting to fix the vehicle with a contact team, towing it to a vehicle collection point, extracting it by heavy lift helicopter, or destroying it in place. Vehicle evacuation will often be extremely difficult due to limited MSRs and canalizing terrain features.

1. Contact Team: After the operator has failed to start his vehicle, a contact team should be dispatched to repair or start the vehicle. The contact team should be task-organized appropriate to the mission. Under cold weather conditions, more personnel, tools, equipment, and portable tents and heaters may be required.
2. Towing: If the rapid fix of a vehicle is not accomplished, it should be towed to a salvage and collection point or maintenance collection point (MCP). *Tow bars must be used.* Chains and cables can easily break loose and allow the disabled

vehicle to slide. Units are responsible for vehicles being towed. Unit vehicles can generally accomplish towing, however, wreckers or tank retrievers may be more appropriate. Exercise care when towing a cold-soaked vehicle. Drive trains may have to be disconnected at extreme cold temperatures to prevent further damage.

3. Extracting Vehicles by Helicopter. Nearly all MAGTF assets can be lifted by the CH-53E. The 5-ton vehicles are not commonly lifted but in remote locations, this might be the only choice for vehicle extraction to an MCP. The inclusion of helicopter support team (HST) personnel or the provision of HST training to the contact teams will allow immediate commencement of the lift. Refer to Section 4012: External Operations.
4. Third and Fourth Echelon Repair: The third and fourth echelon repair should only be performed in established rear areas. Vehicles should be transported to MCPs where selective component exchange can be performed as intermediate maintenance if authorized.
5. Vehicle Destruction: Sometimes vehicles have to be destroyed to clear MSRs. Contact teams must be trained in destruction and provided with appropriate materials (jacks, winches, towing gear or explosives) to accomplish this mission.

Still Pending: MCMWTC is working on a cold weather specific PEB and CLD for a standard MEU EDL.

5007. General Engineering

A. Effects of the Environment:

1. Engineer operations are key to success in cold weather operations. This environment increases manpower and equipment requirements, particularly for construction and maintenance. Engineer operations are influenced by the amount of the individual's exposure, fear of the cold, bulky clothing, and the effect of cold on equipment. The theater, under the grip of severe or wet cold conditions, will favor those best trained and equipped. To be effective, engineer units must train in the cold weather environment to perform the required tasks. Commanders should plan for more breakdowns and other problems related to the cold. (See chapter III for information on combat engineer operations.)
2. Nearly all logistics items will have to be brought in from the outside and/or initially built or developed in theater. Engineering capabilities, both in pioneering and construction, will be inordinately taxed. Engineer support should be planned for initial development of roads, water systems, airfields, fuel dumps, field fortifications, bridges, buildings, communications facilities, utilities, sewage and solid waste disposal.
3. General engineering support may be provided by host nation support or requested from the naval construction force (NCF) by the MAGTF. The first priority of work for NCF elements will support the aviation combat element in airfield development and maintenance. Transportation must be provided by the MAGTF for the NCF.

B. Terrain: The terrain dictates extensive engineer operations in the cold weather operating area. The effects of cold weather on the terrain are discussed in chapters 2 and 3.

1. Tundra: Not all cold weather terrain is tundra. Road building in tundra is difficult, time and material consuming. Bypass is recommended if possible. Tracked and wheeled vehicles must try to use the few existing road nets. Roads or trails that are plowed or graded during winter operations may become impassable under thawing conditions.
2. Rivers and Lakes: Frozen rivers and lakes can be used as winter highways. Engineer units must become familiar with the load carrying capacity of ice and how to determine trafficability. Engineers must understand how to establish sympathetic mine fields on rivers and lakes.
3. Snow and Ice: Well-packed snow and ice has good load carrying capacity for equipment with low ground pressures. However, caution against collapse must be taken when crossing a snow bridge that covers a stream or crevice. If the temperature is below 10 °F, ice bridges can be constructed. Snow tends to level out the terrain and make movement easier. Be careful when parking vehicles on ice fields or ice bridges. The heat from the vehicle may weaken the ice to the point that it gives way.

C. Mobility: In many contingency areas, preplanned engineer capabilities will have a great effect on enemy mobility, canalizing and/or blocking enemy advances. Natural occurring events will inhibit mobility. Ports may be blocked by ice; roads may be ribbons of snow or ice. Snow, rain, ice, and mud may impede, reduce, or shutdown mobility for extended periods. The mission of receiving supplies and moving reinforcements—both men and equipment—will severely test the engine capability of assisting in resupply and CSS. Providing CSS is more difficult due to a lack of road, bridges, rail lines, buildings, water systems, and aids to navigation. Roads, bridges, and ports that do exit are critical and will need to be denied or developed ASAP. More stream crossing equipment, ferries, fixed and floating bridges will be required or need to be destroyed. Cross-country movement will require engineer support. The difficulty of constructing field fortifications is magnified. Quantities of heavy construction equipment must be increased. All the equipment must be winterized. Due to the temperature, deep fording is not a recommended option regardless of the time of year. As a result, there is an increased requirement for manpower, crossing equipment installation, and maintenance. Drainage throughout the sub arctic complicates efficiency because rivers flow north and ice starts to melt in the South. This causes flooding until the river mouths thaw.

1. Mobility/Countermobility. To meet the need of supported units; engineers must be highly mobile and engineer equipment able to transport over ice and snow. Bridging equipment, over-snow vehicle (BV-202/206), and helicopters will be required to move men and equipment, particularly in summer, early winter (freeze-up), and spring (thaw and breakup). Before any operation in a cold weather area, engineers should make a comprehensive map study to determine bridging requirements, especially for crossing streams and lakes, which are not continually frozen solid. Minefields can be developed to deny roads, bridges, airfields, LZs, and ports.
2. Snow and Ice Removal. During heavy snowfalls, snow removal will be necessary to clear main LOCs and airfields. In some theaters, snow and ice removal will be an HNS (host nation support) responsibility, which will require coordination by the CSSE engineer officer. In other theaters this support will need to be requested from the NCR. Normally, snowplows, graders, angle bulldozers, drags, dump trucks, and front-end loaders remove snow. The Marine Corps has no specific snow-clearing vehicle for use during the assault. USMC/USN bulldozers and plows are not armored and are vulnerable to enemy fire. Roads and staging areas may need to be continuously cleared. Road surfaces can be compacted, and then covered with sand, gravel, MO-MAT, or some other material that will increase friction. (See FM 5-101.) In planning a road clearing operation, special consideration needs to be given to collecting and dumping of removed snow. This is important so that existing snow accumulation does not create obstacles and does not impede future removal operations. Engineers will need special training and equipment for snow and ice removal. Commanders will find it necessary to develop special training packages to fill this void. Organic engineer equipment must be modified and the operators need to be proper trained to handle snow removal versus dirt moving operations. This training may be available by cross training with host nation units.

D. Heavy Equipment: In general, the amount of engineer heavy construction equipment must be supplemented/ increased with crawler (tracked) type replacing less mobile wheeled tractors. Tracked personnel and cargo carriers must be added to permit equal mobility of supported and supporting units. Additionally, special purposes equipment (ice augers, portable duct heaters, and extra maintenance shelters) must be added to compensate for the environmental conditions. Engineers must be able to function in temperatures below -25 °F and in long periods of darkness. Heavy equipment operators will need heated cabs and equipment, which have been prepared for cold weather operation. Before embarking, operators should be thoroughly schooled in vehicle operation and maintenance in cold regions.

1. Trafficability. Unfrozen tundra and muskeg pose unique problems to moving wheeled or tracked vehicles. One pass of a tracked vehicle through an area can lead to degradation of the tundra surface leaving a quagmire. This is a similar effect on most frozen ground surfaces.
2. Personnel Considerations. Additional operators and mechanics are a must in cold weather operations. Four operators for each major item of equipment are recommended. Operators rapidly become fatigued and must be relieved after short periods of time. Operators must use the buddy system in the cold, to watch for signs of frostbite on fellow Marines and for fatigue, which might affect safety. Continuous operations, except for short periodic stops for operator checks and minimum equipment maintenance, prevent equipment from freezing. A small amount of heat in even a simple enclosure will pay dividends not only in increased productivity, but will help prevent cold weather injuries. Small portable heated warming tents provide warming areas for contact teams and other maintenance personnel away from established shelters.
3. Maintenance Considerations. Maintenance in cold weather consumes a large portion of the total attention of any force. Cold weather maintenance and servicing procedures must be followed. Factors that increase maintenance requirements include:
    - Long distances over which operations are conducted.
    - Increased heavy strain of cross-country movement on all equipment.
    - Equipment will at times operate continuously 24-hours per day.
    - The general effect of environmental factors in making all activities slower and more difficult.
    - An adequate stock of repair parts proportionate to the longer periods of use per day of all equipment is required.
    - Rapid cooling and re-heating of all mechanical parts quickly increases wear.
4. Rubber-Tired Equipment: More tires will be needed due to higher failure rates. Chains or studded tires are necessary to traverse snowy, icy ground. Snow and ice chains will wear tires much faster than normal. All vehicles should be equipped with proper chain repair kits and chain tighteners. Drivers should be schooled in proper installation methods as well as safe driving techniques and proper movement speeds with chains. If available from host nation support or commercial vendor, 3/8" hexagonal chains perform better than loop chains. Personnel need to be trained on the proper tire chain installation, maintenance and fabrication.
5. Tracked Equipment: Marine Corps tracked vehicles are not specifically designed for operation in cold regions during summer or winter. Special techniques are required to safely and successfully operate this equipment under these conditions.

Special driver/operator training is required. Operators must prevent ice build up in the drive wheels.

E. Special Equipment Requirements: Performance of the mission in the cold will require specialized equipment. This equipment must be planned for. Most items are found in the combat/support engineer's T/E or maintained by C/TEP. Some items have to be specially purchased.

1. Chain Saws. Use chain saws to cut ice, snow blocks, and vegetation for shelters, fortifications, and cover and concealment. Commercially available power drive adapters to chain saws can accept ice augers. Purchase of such kits would place more augers in the hands of Marines needing them.
2. Ice and Snow Saws. Use ice and snow saws to cut frozen snow and ice for fortifications, tunneling, and building troop shelters.
3. Ice Augers and Ice Testing Equipment. Power ice augers are needed to gain access to water and for ice reconnaissance, breaching, and testing. One commercial of the shelf (COTS) item that is versatile in this requirement is a gas-powered auger designed for one or two man use.
4. Steam Generators. Steam generators may be used to thaw construction material as well as aiding in digging fortifications. They may be used to free vehicles, tanks or equipment frozen on the ground's surface, stuck and frozen in the ground, or to remove frozen ice and mud from the tracks of tracked vehicles. Other potential uses, including "steam cutting" in ice and frozen soil, thawing frozen equipment and water and fuel lines, embedded into mud or ice. Additionally, it can assist in the placing of obstacles and mines in frozen materials.
5. Water Pumps and Hoses. Water pumps and hoses will be required in areas where ice bridges are being constructed. Make sure there are adequate heated areas to keep this equipment from freezing during and between uses.
6. Water Buffalo Heaters. Use special heaters to keep water in buffaloes from freezing. Care must be taken not to contaminate the potable water during use. These are required at water points to assure that water is always available. M80 water heaters and immersion heaters work best when used with specially designed stainless steel water buffaloes. Special care and consideration must be taken when using 6-cons for they lack a heating device, therefore they are prone to freezing. **(Water Heater/6-con NSN pending)**

F. Construction: Cold weather operations magnify field construction time and the difficulty of conventional engineer work. Environmental characteristics that complicate engineer tasks are permafrost, extreme and rapid changes in temperature, high winds, snow, ice storms, and flooding.

1. Local Materials: The availability of local materials must be considered when planning construction projects. Local materials may be extremely limited. Where not available, the pace of construction may be considerably slower. Knowledge of local resources (host nation liaison officers) permits ready identification of materials useful for the speedy initiation and completion of construction projects.
   a. Sand, Rock, and Gravel: Location of suitable sources of rock or rock deposits, gravel, and sand for aggregate are essential. The presence of frost or permafrost affects the availability of soils. Gravel reduces the thawing of frozen ground and sinking from melted permafrost. It insulates directly and

reflects sunlight. Filling vehicle ruts and traffic-created mud holes are an effective use of gravel. Gravel is frequently available from present or past glacial moraines. Terrain features often indicative of gravel deposits are streambeds, low hills, and outcropping of trees. Accessibility will often depend on the season. Glacial deposits of sand and gravel may also be found along the seacoast, lakeshores, and the backwaters and meandering channels of rivers.

b. Timber: Trees do not grow significantly in high mountainous regions or above the Arctic Circle. Consequently, the need for timber and building materials must be anticipated. In sub arctic regions, timber is usually available within a range of 50 to 100 miles. If a need is anticipated, it must be planned for and brought in from outside the area of operation.

c. Snow and Ice. In a country where normal resources are limited, natural building materials like snow and ice must not be overlooked (See ch. 3004 *Combat Engineer Operations* and App. B. *Field Work and Camouflage*).

2. Horizontal Construction. Engineers will need specialized training in horizontal construction in the arctic/sub arctic. These problems will be magnified due to the depth of permafrost and/or the freeze-thaw cycle. Engineers will need practical application training in building berms for fuel and ammunition dumps. Try not to disturb frozen ground areas, which might, in a freeze/thaw condition, develop dams that could become areas of standing water. Greater attention and planning are dedicated to area drainage due to potential thawing.

a. Road Construction: Road construction is time-consuming in the summer. Road building requires large quantities of gravel. A thick layer of gravel is needed to keep the permafrost layer from thawing and the road from sinking. Permafrost is not necessarily present in cold weather areas. It may be uniform or discontinuous. This should be determined by engineer reconnaissance. In arctic and sub arctic areas, thick layers of peat may need to be removed before construction can start. During winter months after the ground freezes and snow falls, snow and ice roads and work pads can be built in many areas with a minimum of bulldozer work. Engineers will be called on to evaluate the strengths and capability of ice (See ch. 3004 *Combat Engineer Operations* for ice and ice reconnaissance information.).

b. Rivers used as water highways in the summer and ice roads in the winter significantly enhance movement. Some roads and trails exist but may be undeveloped and limited in trafficability. Road networks and railroads may be practically nonexistent making road construction a major operation. Cross-country movement of units without engineer support is extremely difficult.

c. Airfield Surfacing: Various types of prefabricated mats may be successfully laid over bare, frozen ground, compacted snow, and ice. They are difficult to handle in the cold windy environment. It is more economical to use abundant snow and ice in the construction of temporary winter airfields than to transport mats to remote arctic/sub arctic areas at the beginning of construction operations.

3. Airfield Maintenance.

a. Effect of Bombing. The effect of bombing must be taken into account. Experiments show that a 1,000-pound bomb dropped on sea ice, 4½ feet thick, produces a hole roughly 100 feet in diameter. It scatters blocks of ice up to 5 feet square over an area 180 feet in diameter. These blocks quickly freeze to the surface ice and form a complete obstacle to the landing of aircraft. Several days will elapse before the ice is strong enough to carry equipment necessary to restore the surface.

b. Thawing Temperatures. Runways constructed in winter will become non-operational as soon as thawing temperatures begin to melt the compacted snow surface. They will not be operational any time when the temperature of the air combined with the heat of the sun raises the runway surface to 32 °F or higher as the surface becomes slick.

c. Snow Removal. To provide space for removed snow, leave at least 150 feet open between the near edges of parallel runways and taxiway. Runways to be used by the heavy aircraft must be kept as free as possible of snow and completely free of ice. 24-hour snow removal operations are required to maintain adequate safety in heavy snowfall areas.

d. Effect of Ground Water. Because of ground water, it is generally better to locate the parking aprons and attendant facilities on the side of the runway toward which the permafrost table slopes.

4. Snow Fences: **(Pictures to be added at a later date)**

a. Use. Snow fencing is used to control wind driven snow, in order to prevent the drifting snow from interfering with the trafficability of the main roads. Commercial snow fences are commonly used. They consist of metal posts and wooden laths or metal pickets about 5 feet long woven together with wire. Log and brush snow fences may also be used.

b. Location: Before winter begins, conduct reconnaissance to determine where to place snow fences to control drifting snow from forming obstacles that will impede vehicle or personnel movement.

c. Placement: Snow fences are placed on the windward side of the runway/roadway, as determined from the prevailing winds. Fence height determines the distance it is to be placed from the traveled way (generally 15 times the height of the top of fence). Heavy-duty barriers, similar to snow fencing, can be used to reduce the about of debris and snow from collecting on MSRs in avalanche prone area. **(Picture: Types of Snow Fences, Placement, and Snow Fence Control)**

d. Erection: Snow fences are erected before the ground is frozen. Metal posts are driven into the ground. Fencing is wired to the windward side. In heavy snow, use long posts so that fencing may be raised on the posts as the season progresses. This will increase snow storage to the leeward side. Fencing is initially installed with the bottom about 6 inches above ground level to prevent the ends of pickets from freezing fast. Such freezing makes it difficult to raise the fence and may cause the pickets to break off when swayed by the wind. Brace end posts according to anticipated wind velocities.

5. Bridging: Drainage throughout the arctic/sub arctic is complicated and inefficient. Rivers flow north and the ice starts to melt in the south. This causes overflow, flooding, and a considerable buildup of hydraulic pressure until the river mouths are thawed.
   a. Bridge construction procedures used in temperate areas can be used in the arctic if special precautions are taken. Consideration must be given to construction materials available and the type of vehicle to cross the gap. Over the snow vehicles have less ground pressure per weight than most military vehicles. Special bridging requirements may be brought about by stream characteristics (heavy, fast current), steeper slopes found in mountainous terrain, and changing ice conditions.
   b. Accurate information is necessary for proper planning and construction of bridges, applying stream crossing techniques, and using military bridge equipment in remote arctic and sub arctic areas. Consult with host nation liaison officers. In areas that have been developed and exploited by civilian activities, the local inhabitants and other personnel engaged in developing the area are excellent sources of information for stream and river characteristics. Information must be obtained by:
      - Field reconnaissance
      - Interpretation of aerial photographs
      - Interpretation of arctic geographic factors
   c. The ideal nonstandard bridge design for arctic use, as depicted in FM 5-34, provides minimum restriction to river flow. A design should provide a sufficient height to clear floodwater, flow ice, and debris from ground and stream icing formations. Clear spans are better than those using midstream trestles or crib piers are. Special ice protection is almost always required to prevent excessive damage to any type of supporting structures. Except for these considerations, designs in FM 5-34 and allied text should be followed.
6. Demolition. Demolition (see app. B) will be complicated by the frost or permafrost levels and mud during the seasonal transition periods. Explosives are in great demand for use in construction and for performing tasks, which, in temperate zones, are normally done by machinery.
   a. Cold's Effects on Demolition: Explosives are affected in cold temperatures.
      - Explosives tend to detonate with reduced force.
      - They become brittle and difficult to use.
      - It is difficult to keep non-electric firing systems dry and usable.
      - Static electricity, which easily develops in the cold, dry conditions, makes non-electrical detonation techniques safer than electrical initiated techniques.
      - Extra care must be taken with storage, handling, and preparing explosives.
      - Condensation increases the chances of misfires in the cold. Hang fire and misfire times can double in cold regions.
   b. Cold's Effect on Personnel: Use of demolitions in the cold is hampered by the decreased efficiency of troops placing charges. Preparing and placing charges must be done with bare hands. This limits the amount of time each Marine can work. In extreme cold, handling of explosives becomes a partnership or team

effort, because the demolition team must wear contact gloves. (Note this practice is discouraged in FM 5-25 *Explosives and Demolitions)*

7. Explosive Ordinance Disposal (EOD): Frozen ground or permafrost can reduce ability to excavate and remove unexploded ordinance (UXO) buried in the ground. Severe low temperatures can degrade the equipment performance of the EOD technicians and affect the fusing mechanisms in UXO, perhaps making them more susceptible to detonation. Technicians working in cold weather can reduce there work efficiency because of bulky clothing and an inability to work for long periods of time in the cold; plan for EOD missions to take longer than normal to accomplish. Mountainous terrain will reduce the mobility of teams, thus increasing incident response. Retrieval of disarmed UXO will be more difficult in mountainous terrain; most of UXO will need to be destroyed in place. Irregular terrain features will make sweeping and locating UXO more difficult and time consuming, increasing the time required for tasks.
8. Water Supply. In cold weather, the chief source of water supply for large units in the order of their efficiency and economy are:
   - Drawing water from under river or lake ice.
   - Melting ice and snow.
   - Well-drilling capability is included in Naval Construction Force (NCF) assets. (See NAVFAC 315, *Naval Construction Force Manual,* chapter 6, and *Instructional Publication* 1-4, *Marine Fleet Organization 1990.)*

   a. Snow can be melted for a water source but it is not very efficient and is very time consuming. Large quantities of snow produce small quantities of water while burning large amounts of fuel. If the only source is snow, granular snow is preferred, usually obtainable near the ground, due to it has higher water content than the lighter snow of the surface layers.

   b. Water can also be made potable from a brackish source, like a fjord or seawater, using the reverse osmosis equipment (ROWPU). Water points on lakes and rivers should be located on the leeward side where there is generally clearer water, less snow drifting, and more shelter from the wind. Brackish water must be preheated before being treated in the ROWPU. Additional guidance is provided in FM 5-104.

   c. In extreme cold, heated shelters are required to operate water purification, storage, and distribution systems. Heat must be continuously introduced to the incoming raw water and to the product water in order to prevent the distribution system from freezing.

   d. Holes can also be drilled through ice by the use of hand augers, however, shaped charges are far superior to hand tools in preparing water holes in thick ice since hand tools are generally inefficient if ice is over 60 cm (24") thick. (Ice usually will be thinnest where the most snow covers it.) The methods used, however, vary with the condition of the ice and with the equipment, personnel, and time available.

   e. Ice mining. In arctic the environment, a satisfactory source of water is obtained by gathering ice from a designated mining area, transporting to shelter, and melting it in containers over the shelter heaters. Arctic ice is frozen seawater. Concentrated brine does not freeze and slowly drains from the ice, Therefore first year ice, which is smooth and flat, is salty and not drinkable. The preferred color is a clear blue tint.

f. When transporting water by six-cons and water bulls, consideration must be given to the quantity of water being carried. They should be filled and transported with no more than ¾ full. This is needed for the agitation that the bulk liquid experiences during transport. This movement maintains the water from freezing due to energy constantly being added. Planners need to ensure that time/distance factors are considered when planning supply routes. As the time increases without energy constantly introduced, the less heat the water retains and increases the risk the water would freeze. This process becomes much more rapid as the volume of liquid decreases due to distribution.

9. Hygiene Services:
   a. Bath and Laundry: Bath and laundry units need to be established immediately adjacent to rivers or lakes to reduce water freezing between the sources of supply and the point of use. Consideration must be given to disposing of wastewater. If a moving water source is used, remember the affect of discharging on downstream units. Preventing waste runoff from contaminating a stationary water source is also critical. Wastewater will freeze in the cold close to where it is deposited. During transition periods, this wastewater will melt and move rapidly once the mouths of the rivers (in the north) break up.
   b. Human Waste: Heads will need to be established in permanent bivouac sites. Due to frozen ground/abundant snow cover, digging heads may not be possible. Provisions need to be made for collecting, storing, and disposing human waste in areas where troops will remain for longer than 24 hours. Waste must be burned, hauled away or buried as a last resort.

References:

Coordinating Draft MCWP 3-17. Engineer Operations, sec 4009
Coordinating Draft MCWP 3-17.2 Explosive Ordinance Disposal, sec 4009
FM 31-71, Northern Operations, section VIII, Engineers
FMFM 7-12, Cold Weather Operations, Section V
NAVFAC P-992, Underwater Construction Teams Arctic Operations Manual, June 1994
FM 5-104 General Engineering, Nov 86, HQ, Dept of Army
FM 5-25 Explosives and Demolitions, Nov 86, HQ, Dept of Army

5008. Health Services Considerations.

A. The Mission: The mission of health service support is to provide the medical and dental care to maintain, preserve, and restore the combat power of the force. Inherent in this objective are the requirements to return personnel to duty as expeditiously as possible and to minimize morbidity and mortality in those cannot be returned to duty in a timely manner (FMFM 4-50, *Health Service Support*). Cold or mountainous environments make accomplishing this mission more difficult. Unit commanders/ commanding officers are ultimately responsible for the health of their commands. To assist in carrying out this responsibility each commander is provided health service support, either through organic medical elements or through medical elements of a designated supporting structure. When additional medical support is required, specific requirements become a subject of the planning process. Appropriate sources for the necessary support are identified and tasked.

B. Preventive Medicine: Preventive medicine carries special importance in cold weather environments. Personnel with active medical problems may bring problems upon others as well as themselves. Respiratory infections for example can make entire tent teams ineffective. Another factor of preventive medicine is the screening of individuals at high risk for injury or illness in the cold. Preexisting orthopedic injuries will often worsen under the increased stress of movement over snow. Individuals in poor physical condition are more likely to suffer orthopedic or traumatic injuries requiring time and manpower intensive casualty evacuations (CASEVACs). Individuals with prior cold injuries such as frostbite are at increased risk of future cold injury.

1. Proper supplies are very important in maintaining an effective force. The availability of adequate clothing will lessen environmental injuries such as hypothermia and frostbite. Ensuring personnel use sunscreen and sunglasses will help prevent the severe sunburns possible at high altitude or in snow-covered environments.
2. The MAGTF surgeon is responsible for staff supervision of medical activities including environmental and food service sanitation. Proper sanitation will help prevent the spread of infectious diseases, which have incapacitated armies throughout history. In a cold environment personal hygiene is more difficult to maintain due to limited water and inconvenience of bathing. Potential for spread of infectious diseases is increased by the condensed living spaces, such as tents, shared by several individuals.

C. Health Maintenance: The function of health maintenance includes those tasks required to ensure the medical and dental readiness of the unit and its personnel. A preventive approach must be taken where personnel are screened for active medical problems prior to deployment. Dental records should be screened and all personnel brought to class I status prior to deployment. Those with active medical problems should not be deployed to front line operations, where medical and evacuation assets are more limited. During cold weather operations those who develop communicable infections should be isolated early before others are affected. Commanders at all levels should encourage appropriate water consumption to reduce dehydration, which predisposes to many medical problems.

D. Casualty Collection and Evacuation: Casualty evacuation may be broken down into two phases. The first phase involves movement of the casualty from the forward line of troops (FLOT). The second phase involves movement of casualties from the Battalion Aid Station (BAS) or Regimental Aid Station (RAS) to the rear. Current Marine Corps doctrine calls for medical companies to move forward to pick up casualties. Generally this is accomplished by helicopter or ground transportation. Due to enemy fire, weather conditions, or high altitude, helicopters often cannot fly to the FLOT for casualty evacuation. The terrain may prevent transport by larger motorized vehicles as well. Snowmobiles outfitted to pull litters may provide a quick means of casualty evacuation. If vehicles capable of negotiating snow-covered terrain are not available, unit transport of the casualty by hand may be required. Litter transports are manpower intensive, requiring involvement by individuals other than medical personnel. All personnel in front line units should be familiar with the basic skills necessary for litter transport of casualties over snow or steep terrain. Each company-sized element should include personnel, such as Winter Mountain Leaders, who have experience in route planning and efficient movement techniques. Assistance by those trained individuals may be required to help negotiate terrain obstacles tactically.

1. Injured and immobilized patients are at great risk of cold injury. They must be well insulated during transport. Chemical heat packs and charcoal powered personal heaters should be used to help keep the casualty warm. Lengthy CASEVACs require establishment of aid stations along the evacuation route. At such intermediate stations casualties may be reevaluated and re-warmed. Litter teams may also be rested or replaced with new personnel.
2. Once the BAS or RAS has been reached further casualty stabilization and treatment may take place. Evacuations from this level of care to a larger facility need to occur as rapidly as possible to prevent exhaustion of the BAS' limited treatment capabilities. If available, helicopters are the preferred means of transport due to their speed. Poor or icy roads require the availability of chained or tracked vehicles for ground transport from the BAS or RAS. Both medical facilities should be located close to a viable road when possible.

E. Casualty Treatment: Casualty treatment includes providing the care which is within the unit's capabilities. Tasks include triage and all levels of treatment ranging from self-aid and buddy aid through initial resuscitative care. The adage *there are never enough corpsmen* applies, the shortfall in medical personnel dictates the services of corpsmen are optimized through triage. This may require initial treatment of certain injuries by non-medical personnel. All personnel should be trained in the self-application of tourniquets or pressure bandages to control potentially life threatening bleeding. The lengthy CASEVACs seen in snow-covered regions require the forward movement of treatment and medical supplies. Supplies such as IV fluids will need to be spread loaded among the troops.

1. Environmental injuries such as hypothermia, frostbite and trench foot have played deciding roles in military operations throughout history. The impact of these injuries may be greatly decreased by ensuring personnel have proper food, water and clothing. Food and water requirements increase significantly in cold environments. Personnel must be educated on proper use and maintenance of cold weather clothing and footwear systems.
2. Cases of severe hypothermia, with a core temp of less than 90 degrees Fahrenheit require specialized patient transport. Steps should be taken to prevent further heat loss. Evacuation without rough handling, similar to that used in spinal injuries, must be employed to lessen the chance of a lethal cardiac disorder.
3. Frostbite will require the removal of the affected individual from further exposure to the cold. Frostbite should be re-warmed in a specific manner requiring medical equipment normally unavailable in the field. The part should not be re-warmed if additional exposure to the cold with re-freezing of the part may occur. For these reasons frostbite should be evacuated directly to a treatment facility in the rear as soon as possible.
4. Heat injuries may also occur in cold environments. They are most common when dehydrated individuals over dress during periods of high exertion. High levels of exertion in cold or high altitude environments lead to large evaporative fluid losses. The importance of removing layers of clothing or venting during periods of high exertion must be stressed. Air in cold or high altitude environments is often very dry contributing to increased respiratory fluid losses as well. Without frequent encouragement Marines will not consume enough water to maintain an adequate fluid balance.
5. Many areas of the world where cold weather operations occur are also at high elevation. There are medical illnesses ranging from mild to life threatening associated with exposure to high altitudes. The body must go through a series of adjustments for acclimatization to the decreased barometric level of oxygen available at high altitudes. Staged ascent of 1,000ft to 3,000ft a day above 7,000ft in elevation allows for more successful acclimatization. If an individual fails to acclimatize, they may experience acute mountain sickness (AMS) which is marked by headache, nausea and vomiting. AMS is rare below 6,500ft. This mild disorder will improve if further ascent is avoided. However, an individual may develop life threatening High Altitude Cerebral Edema (HACE) or High Altitude Pulmonary Edema (HAPE) if they ascend further with AMS. Both disorders are rare below 10,000ft in elevation. Both HAPE and HACE require

immediate medical treatment and descent or death may occur. Medications such as diamox may be used to assist acclimatization when the mission requires a rapid ascent. Medical personnel should be well versed in the identification and treatment of high altitude disorders if deployed in high altitude, mountainous regions.

F. Training: Due to the increased demands on the limited medical personnel during mountainous operations Marines must receive increased training. They should be able to recognize and initiate treatment of medical conditions ranging from trauma to environmental injury. Successful training will cover the measures taken to prevent injuries. Several members of each Battalion should receive advanced mountaineering training through courses such as the Mountain Leaders School. Those individuals may be needed to provide technical assistance on CASEVACs. Medical personnel should be proficient in all aspects of cold weather medicine. All personnel should be familiarized with the litters used for over snow and steep terrain casualty evacuation. Small unit leaders should also ensure that their personnel are proficient in converting the team sled used for gear transport into a patient litter for over snow evacuation.

G. Medical Supplies. Medical supplies will need special attention throughout the entire logistics chain. Many supplies that are water-based liquids will degrade if exposed to below freezing conditions. If there is the potential for freezing prior to usage they will require special packaging and handling. The AMAL of medical supplies for support of Marine combat operations is prescribed by the MCO 6700.2 series. It gives the number of component kits for different size forces. Specific class VIII (medical) supply requirements must be determined in advance and pre-deployment limited technical inspections (LTIs) conducted. Cognizant medical staff personnel and deploying Marine unit staff personnel must jointly make composition determinations and advance inspections. An adage applies here: if it hasn't been checked, it isn't there. The supplementary AMAL 685 should be requested for cold weather operations. It contains additional items useful in the treatment of medical conditions common in cold weather operations. It is recommended the standard AMAL be augmented with additional supplies needed in the treatment of orthopedic injuries. Useful medications for the treatment of high altitude illnesses not standard in the AMAL are diamox, decadron and nifidipine. Sunscreen should be available either through medical or unit supplies. **(Additional AMAL requirements pending)**

H. Special Equipment: If the terrain requires patient movement over steep slopes, cliffs or ravines, litters capable of withstanding the stress of such transport should be available. Mountaineering equipment such as ropes, carabiners, prussics, and figure 8's should be available for usage in CASEVAC. Tracked vehicles or HMMWVs equipped with chains will need to be reconfigured for use as ambulances in order to negotiate snow and ice covered roads. Snowmobiles equipped with patient evacuations sleds may be useful in moving patients from forward areas to the BAS or viable road. A Gamow bag is recommended when operations at elevations greater than 14,000 feet are planned. Gamow bags are portable hyperbaric chambers used in the temporary treatment of HACE or HAPE. They may be obtained from civilian medical suppliers.

5009. Messing Considerations.

One of the only amenities that remain for forces in the cold weather warfighting situation is subsistence.

A. General: Cold weather subsistence support is not currently covered in MOS schools for 3381 cooks. Nor is it practiced, for the most part, in any training environment except the Norway exercises such as Battle Griffen.

B. Equipment. Adequate tentage in the cold weather environment is mandatory in order to protect perishables.

C. Types of Subsistence: Three types of subsistence will be addressed:

- A Rations: There is no adequate replacement for hot, fresh food as frequently as possible.
- Tray (T) rations: This is an improvement over the packaged operational ration. It is recommend not to use in excess of 30 days.
- Packaged Operational Rations (POR). All are familiar with the Meal-Ready-to Eat (MRE).

Still Pending: MCMWTC is in the process of obtaining cold weather/mountainous subsistence doctrine from the USMC subsistence school at Ft Lee. It will be included in the final draft of this document.

5010. Rear Area Security (RAS) Considerations.

The rear area for any particular command is the area extending forward from its rear boundary to the rear of the area of responsibility of the next lower level of command. The cold weather environment adds further difficulty to managing a rear area for three reasons:

- The volume of supplies required to support operations in a cold weather environment multiplies the size of the rear area to accommodate bulk storage (Class I and III in particular).
- Management of a traffic plan in a rear area becomes more critical because ice and snow limit trafficability and a traffic accident in the constrictive traffic plan will severely hamper if not impede resupply operations.
- Displacement of the rear area due to threat or operational requirement is extremely time consuming again due to volume, limited trafficability, and limited motor transport assets that can both move bulk stores and survive difficult road conditions.

A. Responsibilities: Security of a rear area such as an Advanced Logistics Operations Center (ALOC) is the responsibility of the local rear area Commander.

1. Staffing: The Operational Commander will obviously commit as many forces possible to the forward effort. The RAS commander will staff the reaction force with Combat Service Support personnel. Because the personnel available to the RAS Commander are those that have other CSS missions (messing, motor transport, communications), the actual RAS staff members will change from day to day. The RAS Commander would be wise to assign at least a force leader and sector leaders who have no other job other than security so continuity in the force can be maintained.
2. Division of Sectors: Commonly, the rear area is divided into Sectors for security. Cohesion of the sectors is essential in order to maintain perimeter integrity. The most difficult task herein, because the personnel actually performing RAS are different from day to day, is actually maintaining a uniform plan with a continuous method of performing the task. Again, at least a skeleton cadre of security NCO's who do nothing else but RAS will be the best way to maintain standard procedures.
3. Proportionality of Effort: Tactical schools often teach, "there is no rear area." However, the RAS Commander must balance the enthusiasm to have a militarily "bombproof" compound with the knowledge that the CSS mission must be performed if the operational forces are to be successful. Consideration must be given to the strength of the enemy, the enemy's last known proximity, and the ability of the Operational Commander to respond to a rear area breakthrough. Also critical is for the RAS Commander to make an educated judgment call on the level of alertness. The members of the security force will most likely be off-shift CSS personnel who need rest for mission accomplishment. The minimum level of alertness is critical in keeping the CSS operation moving.
4. Mutual Support: Commitment of a reserve or "reaction force" is always the last ditch effort of the RAS Commander. In the cold weather environment, the

personnel shortage in the rear will be exacerbated by a larger rear area and a huge number of personnel loitering in the rear area because of "cold injuries" hoping for a good excuse to go inside a tent or structure. The most economical method of actually having a reaction force is to coordinate with other local RAS Commanders for mutual support, rotating the duty between as many local RAS Commanders as are geographically able to support. Rehearsal of the reaction force between rear areas is essential to understand the time requirements.

B. Considerations:

1. Personnel: The biggest enemy of RAS is fatigue. In the cold weather environment, shorter more frequent shifts with coordination between the RAS Commander and each CSS Officer-in-Charge is the only way to keep a rested, alert force.
2. Environment: Tactical considerations for RAS in the cold weather, including fire support, are not substantially different than considerations for any rear area. However, the things that make a rear area vulnerable in a regular environment, are simply bigger signatures in a cold weather environment and must be considered. Examples are:

- Vehicle tracks can be seen from the air
- Messing facilities generate huge quantities of steam
- Sound carries much further over snow than ground and maintenance functions are extremely noisy.

C. Recommendations: The security and displacement of a MAGTF size ALOC must be rehearsed with aggressing forces. CSS personnel must understand that a displacement could actually happen, there is no guarantee of a static ALOC. In this event, CSS cannot stop; the CSS Commander must simply push easily predicted requirements to operational units knowing that Classes I and V are likely to be short. The site selection of bulk stores is critical to flexibility. Consider the following recommendations:

- If possible, situate the ALOC so that bulk stores can be accessed and distributed from platforms that can be moved by a means other than vehicular travel. Example: If bulk stores can be opened and accessed while still mobile loaded on an accessory track of a railhead, keep some of the bulk stores on the flatbed cars. This way they could be temporarily displaced for security during a rear area threat.
- As the proximity to a known threat increases, the supplies must have greater dispersion. Common items are needed by almost all units (class I, III, V, radio batteries and medical supplies). Prepacked quadcons with known requirements that can quickly dispatched to using units should be staged in multiple areas of debarkation.
- Stay away from fixed structures. The more accustomed the operating forces become to the "rear area" always being in the same place, the more fluidity and flexibility the CSS units lose. Also, the longer a CSS unit stays in one place, the more personnel it will accumulate, making security and displacement an even more difficult task.

D. Summary: The keys to rear area security are sound planning, early warning, continuous operational security, tactical deception, proper dispersion, cover and concealment, tactical training of rear area personnel and rapid deployment of sufficient forces and resources to counter the threat. The overriding reality is this: For operational forces to survive in a cold weather environment, their bulk stores must be protected. This will require the attention of the operational commander. Local rear area commanders must understand the fundamentals and principles of RAS for any theater and creatively distribute critical supplies so that they are accessible to forward forces.

Appendix A

Avalanche Danger, Recognition, and Rescue

1. Avalanche Danger

Predisposing conditions (Elevation, slope, prevailing winds, and a lack of vegetation) are constantly found in mountainous cold weather regions. When these conditions combine with the development of local environmental conditions in the right sequence, avalanche dangers exists. Generally some type of trigger whether it is natural or artificial will start an avalanche.

a. Operational Level. At the operational level, the MAGTF commander, subordinate level commanders, and operations officers must be able to recognize when an avalanche hazard is developing in their area of operations. With resources like avalanche maps and advice from natives, they can compare avalanche danger levels based on current and/or forecasted avalanche dangers with mission requirements.

b. Tactical Level. Unit leaders (Patrol leaders through battalion commanders) will often be in an avalanche danger area when or as the danger develops. They need to know how to recognize avalanche dangers in their area of operations; conduct the basic test that will help determine the level of danger; and how to select the safest routes to avoid dangers.

c. All Marines. All Marines need to know how to conduct avalanche search and rescue procedures. They should also all be trained in route selection techniques.

2. Predisposing Conditions (Constants)

Certain relatively constant conditions are found in mountainous regions that predispose annual avalanche dangers. This includes elevation, slope, annual weather patterns, and lack of vegetation.

a. Elevation. A mountain is a landmass that projects above its surroundings. For military purposes, these elevations will range from less than 4,000 feet in maritime areas to over 14,000 in high alpine regions.

b. Slope. Slopes between 15 and 60 degrees are predisposed avalanche danger. They provide a steep enough surface for the snow pack to slide.

c. Annual Weather Patterns. The jet stream (Upper airflow's) moves larger air masses (Fronts) that gather moisture as they pass over oceans (Hot air rises). As they pass over mountainous terrain, they cool and dry out depositing precipitation as rain, sleet, or snow.

d. Vegetation. Altitude and latitude affect vegetation. As altitude and/or latitude increase, the amount of quality and vegetation decreases. Toward the Polar Regions this phenomena of minimal vegetation occurs at lower elevations. A lack of vegetation decidedly affects the development of avalanche danger. Without established vegetation there is little to anchor the snow pack

3. Local Environmental Variation

Local environmental variations affect the metamorphism of the snow pack. These include temperature changes, wind flow, and precipitation. Snow crystals that make up the snow pack are constantly in a state of change (Metamorphism). To evaluate the strength and stability of the snow pack, these changes and their constant interactions must always be considered.

a. Dynamic Snow pack. The snow pack constantly changes due to the effects of temperature, pressure, wind, and the arrival of new precipitation. Study of this constant change (Metamorphism) is beyond the scope of this appendix. However, the important thing to remember is that physical changes enable the snow molecules to develop layers in the snow pack that have very poor bonding capability with the rest of the snow pack. This predisposes the snow pack with the potential to slide (Avalanche). The temperature of the snow layers determines this rate of change. Possibilities of change are endless. Some typical examples are:

- Cold rainfall can freeze on the surface and provide a slippery surface for later arriving snowfall.

- Warm rainfall can percolate through, affecting the metamorphism of layers below the surface.

- Sleet can arrive and develop into a ball bearing type of layer.

- Sun can affect the snow's surface (Producing a glaze) or it can radiate through (Affecting layers below the surface).

- Wind constantly moves the snow around resulting in a break up of the snow molecule and deposition of snow in various places, building dangerous cornices.

b. Temperature. The mountains force prevailing air patterns to rise (windward) and fall (on the leeward). As air falls, it cools rapidly, producing precipitation over the lee slopes. Seasons affect the temperature. With the coming of winter, the sun rotates toward the equator. An absence of or less sun cools the air. Cooling to a large degree determines whether precipitation will fall as snow, sleet, or rain and greatly affects the moisture content of snowfall. Moisture content of snow in turn affects its ability to cling to or bond with the snow pack. A thermometer is an essential item that should be carried when operating in the mountains. Temperature observation during a storm is vital and will indicate that:

- A snowfall that starts cold and then warms indicates a poor bonding surface beneath the surface. (The colder snow will not support / bond with the warmer, heavier snow found on the top layers).

- Warmer temperatures after a storm (Usually above 32 degrees F) will promote an equi-temperature metamorphism. Indicators of rapid warming including sunballing.

- Colder temperatures after a storm (Usually below 32 degrees F) will increase freewater in the snow pack, adding extra weight. This overloading is dangerous.

- The longer a thaw persists, the higher the avalanche danger. Small sloughs will be common. Larger avalanches (Including the whole snow pack) down to the ground, called ground avalanches, also occur.

c. Wind-flow. Wind-flow (The Architect of Avalanches) is a very important consideration in the development of avalanches in mountain terrain (See fig. A-1). Wind determines how and where snow will be deposited during and after storms. During and after a snowstorm, the wind will redistribute the snow. Snow tends to accumulate in localized disposition zones instead of falling evenly. The deepest accumulation occurs in bowls and gullies where the snow may be several meters deeper than in surrounding areas. A lee area may get wind-transported snow from more than one direction, causing larger depositions.

(1) Terrain obstacles interfere with the typical smooth flow of free air. Winds usually align with the direction of the canyon floor. Deflection can drastically change snow dispositions.

(2) Winds will accelerate up over the crest of a ridge from the windward side picking up snow and decelerates down the leeward side depositing snow. This pattern is responsible for cornice formation (See fig. A-2) on the lee sides of ridges.

(3) The wind has great effect on the snow crystals. Due to pulverization, the average size of the blowing snow particles may be only $1/10^{th}$ the size of those that fall undisturbed. Because the small particle-sized, wind-deposited snow crystals are two to four times denser than snow that falls with no wind or in a sheltered area, wind-transported snow quickly takes on a firm slab-like structure. The small particle-sized crystals bond with many contact points.

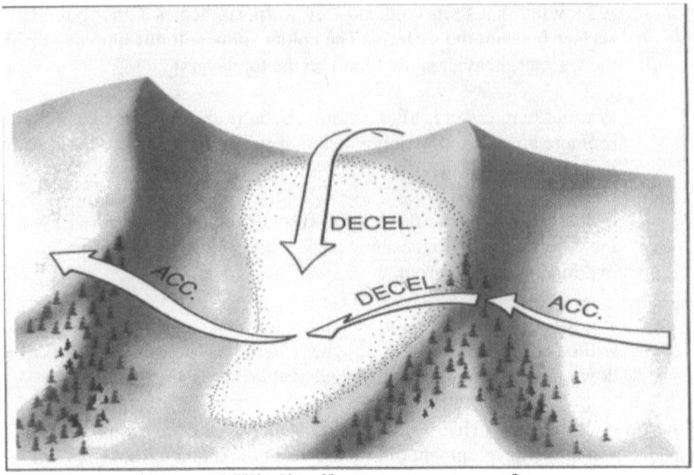

Figure A-1 Wind's effect on movement of snow.

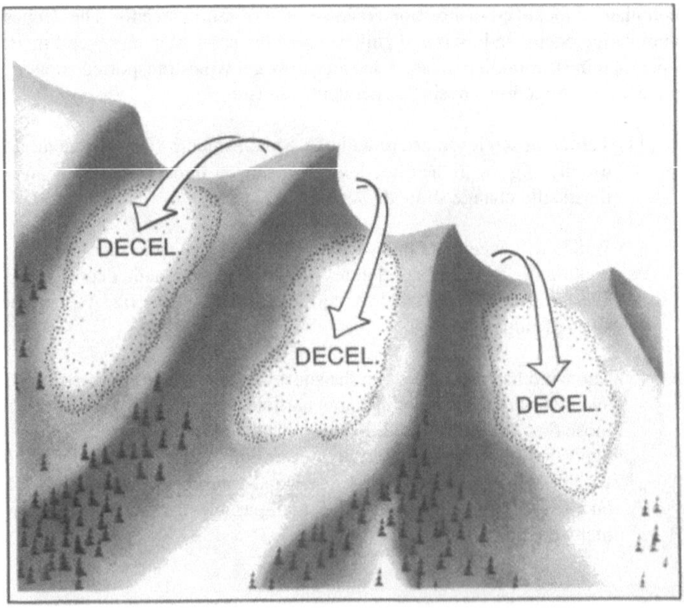

(4) Large amounts of snow accumulate and form cornices wherever the terrain bends sharply. The most common places for cornices to form are on the leeward sides of ridge crests and gullies. Cornices (See fig. A-2) grow as

successive layers and is added during each period of snow transport. The snow deposited on the top overhanging portion is called the roof. After attaching to the roof, layers are deformed slowly by gravity and bend toward the cornice face like a curved tongue. Throughout a cornice's life, the entire cornice will deform steadily outward over the slope below, usually reaching a precarious balance. Cornices often extend as much as 15 meters upward and outward from a ridge crest, depending on ridge shape and steepness of the leeward slope below. The steeper the leeward slope, the less support is offered to the cornice. Winds in excess of 15 mph for extended periods can create an extended avalanche hazard.

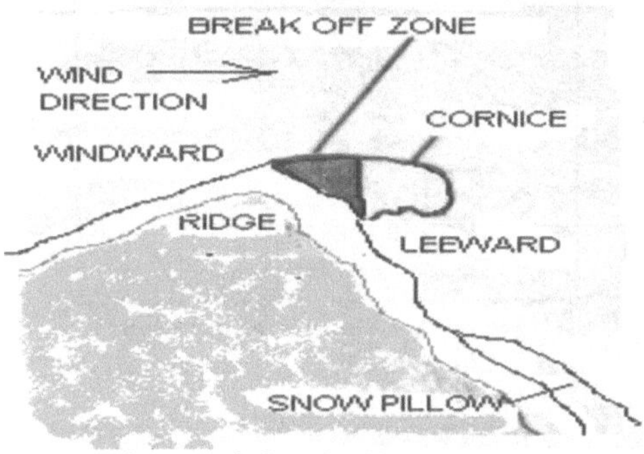

Figure A-2 Cornice Nomenclature

d. <u>Precipitation.</u> The amount of precipitation in the mountains is greater than that found in surrounding lowlands, due to orographic uplift and the temperature characteristics of the mountains. A frontal system moving into a mountain range will be forced to pass up and over the mountain range. As the front moves up, it will drop its moisture. Warm air holds more moisture than colder air. As a front rises, it cools and drops its precipitation, usually slightly windward of the topographical crest. Mountain temperature characteristics will also influence precipitation. Snow characteristics are greatly influenced by temperature. As warm air rises, it cools and follows a certain cooling rate called lapse rate. The average lapse rate will depend on the climate your operating in, and whether it's a dry or wet climate. A good rule of thumb is a 5 degrees decrease for every 1,000 feet of elevation gained. Knowing the lapse rate and the elevation enables the meteorologist to estimate what snowfall temperatures will be at higher elevations.

4. Snowpack Analysis

The key in analyzing a vertical snowpack is in knowing what characteristics exert tension and where the tension exists. These characteristics require an understanding of the concave and convex angles of the snowpack, slope angle, and creep and glide.

a. Concave / Convex. A slope that is convex will put extra tension on the snowpack at the point of convexity. Concave areas will help compress the snowpack in the concave region making the slope much more stable than a convex slope (See fig. A-3).

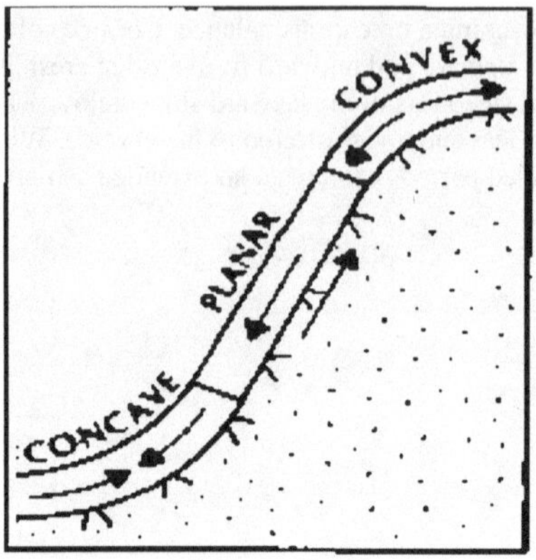

Figure A-3 Slope Configuration Conducive to an Avalanche

b. Slope Angles. All factors being the same, slope angles between 30 and 45 degrees are most prone to avalanche (See fig. A-4). Slopes below 15 degrees will very rarely exert enough tension to cause them to avalanche. However, avalanches have occurred on slopes as slight as 10 degrees, resulting in death. Slopes above 60 degrees very rarely hold enough snow to be an avalanche hazard. However, in areas like the Himalayas, climatic conditions enable the snow to stay on slopes that are this steep.

c. Creep / Glide. Creep is the internal settling the snowpack in response to stabilization. Glide is the snowpack moving (Gliding) down the slope in relation to the ground. These conditions may occur in combination resulting in stabilization of the inclined snowpack over an extended period of time. In larger snowpacks, however, this movement has twisted and even wrecked, strong reinforced structures.

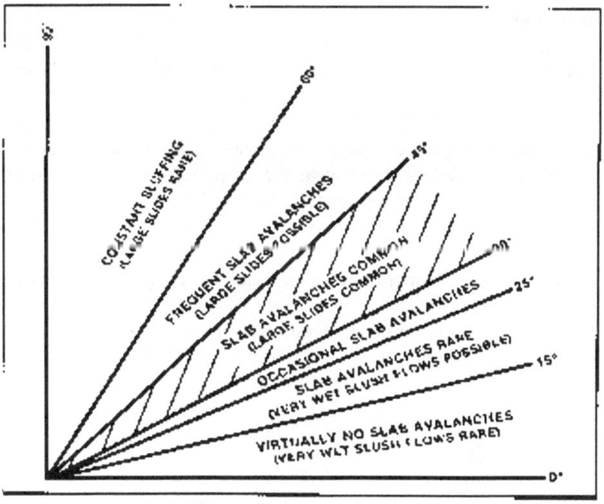

Figure A-4 Slope Angles

5. Storm Analysis

Ninety percent of all avalanches occur during a storm or within 24 hours after a storm. The most important factors in avalanche danger evaluation are wind speed and direction, precipitation and temperature.

a. Wind speed and direction. Wind speed and direction indicate which slopes are being loaded. In light winds, the whole leeward slope is loaded. In strong winds, certain deposition pockets will be loaded rather than being spread out on an entire leeward slope.

b. Precipitation. Snowfall at a rate of 1 inch an hour over a 24-hour period is considered a positive indicator of avalanche danger. Rainfall followed by falling temperatures predisposes the snowpack for follow-on problems. Rainfall on top of a snowpack can significantly overload the snowpack, resulting in avalanches.

c. Temperature Changes. See paragraph 9e.

6. Types of Avalanches

Loose-slab (Loose-snow), slab, and ice avalanches are all dangerous and can result in death.

a. Loose-Slab (Loose-Snow) Avalanches. In the loose-slab or loose-snow avalanche, failure begins near the snow surface. A small amount of cohesionless snow slips out of place and starts down the slope. The initial mass may set an increasingly amount of snow in motion. The avalanche seems to start at a point and fan out, looking V-shaped (See fig. A-5). Loose slab avalanches occur as either wet or dry avalanches and can occur in all shapes and sizes. Small ones are called sloughs and are the most common. Important considerations of loose snow avalanches are:

Figure A-5 Loose-Snow Avalanche

- Although small, they can bury Marines resulting in injury or death.

- During storms, sloughing continually removes snow where the slope is steeper than 50 degrees from upper slopes. This prevents the dangerous buildup of thick layers of snow on such steep slopes.

- In areas where high angle slopes empty down onto lower slopes, sloughing from higher slopes may force the lower slopes to shed small avalanches and become more stable.

- Snow sloughing form upper slopes may overload the lower slopes, creating a greater avalanche danger by triggering slab avalanches.

- Wet loose snow avalanches can have tremendous size if the slopes below the starting points are also wet snow. This is common in spring during a Daytime Melt and Night-time Freeze of the snowpack.

- Dry loose-snow avalanches are common in colder temperatures during Temperature Gradient metamorphism. An extremely powerful air blast may precede the avalanche.

b. Slab Avalanches. Slab avalanches occur when one or more layers of cohesive snow break away as a unit at the fracture line. Slab avalanches occur in sizes from just a few inches to many tens of feet thick and similarly range in width from a few yards to over a mile. Slab material is also highly variable. Slabs may be hard or soft, wet or dry. As the slabs travel down slope, they break into smaller blocks or clods. (See fig. A-6)

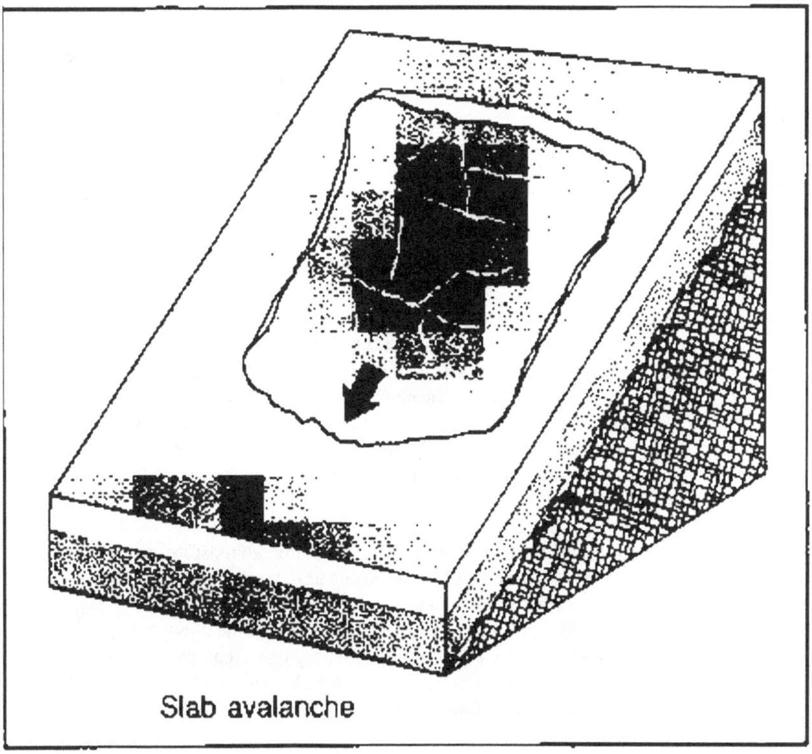

Figure A-6 Slab Avalanche

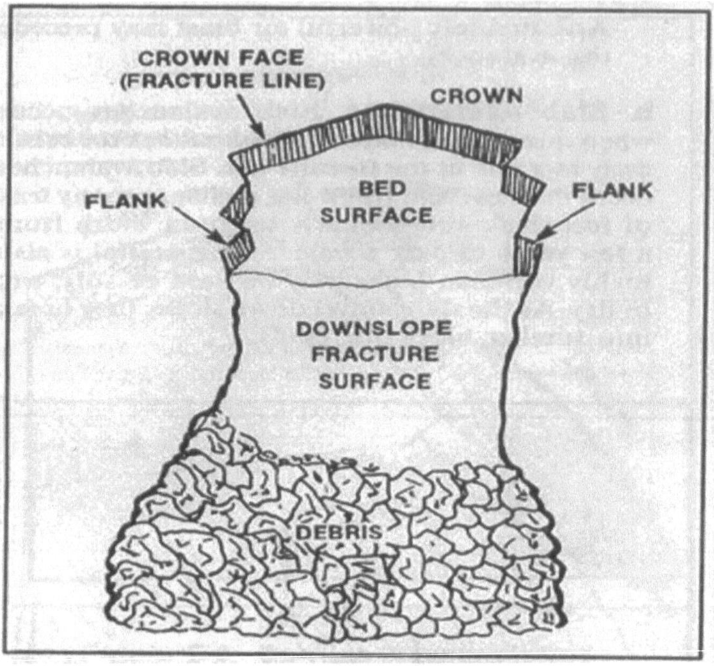

Figure A-7 Slab Avalanche

(1) Origin. Slab avalanches begin with the failure of a slope to hold a slab-like region of snow. The area of the slab will break into small cohesive blocks as it falls down the slope. For a slab to detach completely from the slope, a fracture must proceed around the entire slab (See fig. A-7). Slopes of less than 30 degrees are less likely because there is not enough tension on the slab areas. Slopes of greater than 45 degrees usually slough before slab avalanches occur.

(2) Slab Hardness. The wind is probably the most influencing factor in making slab conditions. Aging and compaction by settlement also contribute. Hard slabs are so hard that they cannot be penetrated by a ski edge. They can survive a lengthy slide down a slope without breaking up. (This indicates a very dense compact slab.) Soft slabs are not compacted as strong. Blocks will break into smaller lumps quickly. The final condition is similar to powder snow.

c. Ice Avalanches. The main causes of ice avalanches are glacial movement and internal melting. Ice avalanches are caused by the collapse of unstable ice blocks (Seracs) from a steep or overhanging part of a glacier. These hanging glaciers are usually easy to recognize, but ice avalanches are generally unpredictable because imminent icefall cannot readily be detected. Unlike snow avalanches that are usually triggered by their victims, ice avalanches usually result from a natural fall.

7. Avalanche Triggers

a. Natural Triggers. Caused by mother nature

(1) Overloading. Too much weight on an unstable snowpack may cause it to avalanche. Overloading may occur from-

- An accumulation of too much new snow too quickly. A general rule is danger exists when snow falls in excess of 1inch per hour within 24 hours.

- The added weight from another avalanche.

- The added weight from a falling cornice.

- Wind-loading of slopes.

b. Artificial Trigger. Caused by man or his equipment.

(1) Man. The added of weight of a skier may cause an avalanche.

(2) Equipment. The added weight of a snowmobile may cause an avalanche.

(3) Explosives. Used by man to control avalanches. Either to keep an area safe or close an area off.

8. Avalanche Path

Once set in motion, avalanches can travel great distances. The path they follow will vary with type and terrain. The average slope angle for avalanche paths that extend for long distances is between 20 and 35 degrees. An avalanche path is divided into the starting zone, the avalanche track, and the runout zone (See fig. A-8)

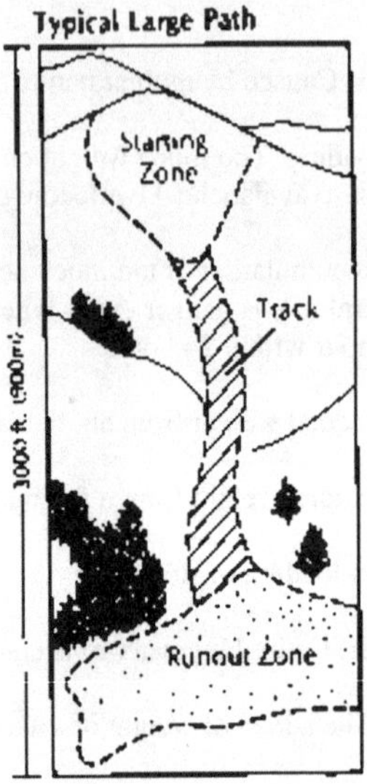

Figure A-8 Avalanche Path

a. Starting Zone. The starting zone is usually steeper than 30 degrees and must have received large amounts of snow. Generally, leeward dispositions, gullies, and bowls are especially big disposition areas. When large amounts of snow are delivered with little or no wind, inactive zones may become active. Most starting zones are bare of trees but dense timber stands can be found in the starting zone if other conditions are right.

b. Avalanche Track. The avalanche track refers to the path below the starting area and above the runnout zone. Avalanche tracks can be channeled or unchanneled. Channeled tracks are confined areas such as gullies and couloirs. Unconfined tracks are on open slopes. Some may have trees present. An avalanche track may be multi-branched and several small starting zones may feed into one avalanche track. It is important to remember that multi-branch tracks may run several times in quick succession. Numerous rescuers have been killed when working in a debris area and a second avalanche has run down the avalanche track within hours of the first avalanche. Wet snow avalanches tend to track a boundary while dry snow avalanches tend to easily jump terrain barriers.

c. Runout Zones. Runout zones consist of the pile up of debris in the area at the bottom of the avalanche track. Variations in weather pattern from one year to another will influence the position of runout zones.

d. Influence on the Avalanche Path.

(1) Avalanche Cycles. Some areas that are normally safe may become avalanche runout zones every 25, 50, or 100 years. Short-term reports declaring an area safe have resulted in loss of life and property because they were inadvisably disregarded.

(2) Air Blasts. Air blasts caused by moving snow preceding the avalanche may extend well past the runout and prove to be extremely dangerous.

(3) Seasonal Changes. Once an avalanche path has begun to slide in a season, it will probably continue to slide throughout the season. This is probably due to the right combination of conditions existing beneath the snow

9. Identification of Avalanche Danger Areas

The best source of avalanche information in a given area is from indigenous personnel. Be aware of short-term observations; i.e., less than 10 years! Most observations of avalanche terrain rely on the following:

a. Slope Angle. See paragraph 4b. Slope angles between 30 and 45 degrees tend to be the most prone to avalanche. Slopes above 60 degrees rarely hold enough snow to be an avalanche hazard.

b. Slope Orientation. It is important to determine how the slope is facing in relation to recent winds and the sun. Subtle changes in orientation (Aspect) can greatly affect snow stability. Be suspicious of leeward slopes because of the additional stress exerted by wind-loaded snow. Be conscious of gullies and draws that may have become loaded with snow during periods of reduced sunlight.

c. Slope Configuration. Avalanches can happen on any snow-covered slope steep enough to slide. On convex slopes, slabs are most likely to fracture just below the bulge where stresses are greatest. On broad, smooth (planner) slopes, avalanches can happen anywhere. Slabs often fracture below cliff bands. Concave slopes provide a certain amount of support through compression at the base of the hollow, but they can still produce an avalanche, especially on large slopes. (See paragraph 4a.)

d. Vegetation. Observations of vegetation will indicate the avalanche history of an area for such things as avalanche paths and runout zone boundaries. Open slopes generally indicate a greater danger than ones with tree cover. Avalanches have occurred on slopes that are tree-covered but this is rare. A simple rule to follow is that if trees are spaced within a few meters of one another (close enough to make ski movement difficult) they will probably provide the necessary protection that makes movement safe. The most convincing evidence of past avalanche activity is a patch of fallen trees, aligned in the same direction, sheared off at about the same height above the ground. Flag trees (trees void of branches on the uphill side) and small islands of trees on an avalanche path are vegetation variations that are usually found high on the slide path. Pioneer growth in a climax forest (trees 1 to 15 years old growing in with trees 100 to 200 years old) and cleared strips of trees in dense forest area are a positive indication of past avalanches.

e. Elevation. Temperature, wind, and precipitation often vary significantly with elevation. Common differences are rain at lower elevations, and snow at higher elevation. There will also be a difference in precipitation amounts, or wind-speed with elevation. Never assume that conditions on a slope at a particular elevation reflect those of a slope at a different elevation.

f. Valley Configuration. A safe travel route can generally be found somewhere in a wide U-shaped valley. Avoid narrow V-shaped valleys. In V-shaped valleys, avalanches can run from either side and continue to run up the opposite side, so there may be little or no safe ground.

10. Indicators of Snowpack Instability

Indicators of instability include-

- Recent avalanche activity on similar slopes and terrain.

- Small avalanches underfoot and visible cracks shooting out form underfoot.

- Booming caused by the audible collapse of snow layers.

- Sloughing debris which is visible evidence of avalanche activity occurring.

- Sunballing (Snow spontaneously gathering into snowballs as snow rolls down hill) caused by rapid warming of the snow surface.

- Dangerous weather patterns to include:

    - Heavy amounts of snow in short periods of time (1 inch per hour in less than 24 hours).

    - Heavy rain that warms and weakens the snowpack.

    - Significant wind-loading that causes leeward slopes to possibly become overloaded.

    - Cold, clear, calm periods followed by heavy precipitation or wind-loading.

    - Rapid temperature rise to near or above freezing after long cold periods.

    - Prolonged periods of above freezing temperatures, for more than 24 hours.

- Snow temperatures remaining cold (Equal to or less than 25 degrees) slow the settlement or strengthening process, thus allowing unstable snow conditions to often persist longer.

11. Indicators of Snow Stability

Indicators of stability include-

a. Snow cones or settlement cones that form around trees and other obstacles, indicating that the snow around the object is settling.

b. *Creep*, the internal deformation of the snowpack, and *glide* the slippage of the snowpack with respect to the ground. (Evidence of these two properties on the snowpack is a ripple effect at the bottom of a slope. They indicate that the snow is bonding and gaining strength through this settlement process.)

c. Absence of wind during storms as indicated by snow accumulation in the trees.

d. Snow temperatures remaining between 25 and 32 degrees ordinarily settle rapidly, becoming denser and stronger.

12. Tests of Snowpack Stability

To finalize a decision on the degree of avalanche hazard present, it is essential to have a method of testing the snow for stability and instability. Described below are the common tests that Marines should know.

a. **Rutschblock (Shear Block) Test**. The Rutschblock test allows trained Marines to assess an avalanche hazard in a short period of time (Usually 20-25 minutes). (Marines conducting the test should be trained as Mountain Leaders by the Marine Corps Mountain Warfare Training Center, British Royal Marines, or by Norwegian schools.) One test does not tell the whole story. Observations from different tests must be integrated with other snowpack indicators, weather and terrain information, and advice obtained from natives. To conduct the test-

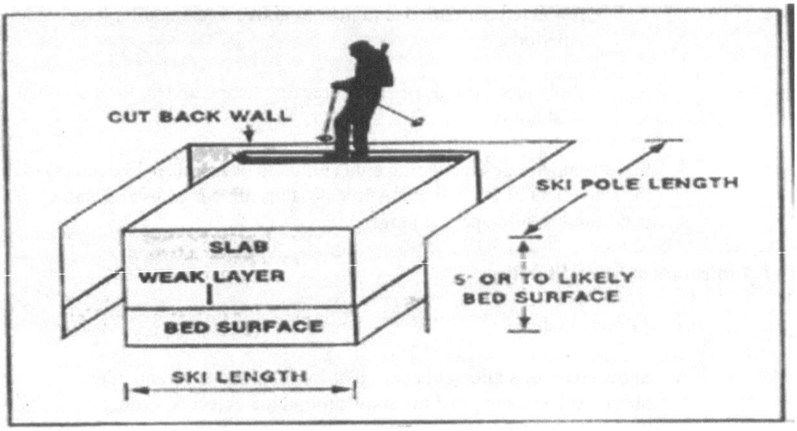

Figure A-9 Ruthschblock Test

(1) Dig the Rutschblock in the start zone of a possible avalanche path. Slope angle must be at least 30 degrees. Be sure that the Marines conducting the test have a safe means of entering and exiting the test site.

(2) See figure A-9 for test preparations.

- Dig two trenches approximately 2 feet wide straight into the slope. These two trenches should be one ski length apart and 5 feet deep or dug to the likely bed surface.

- Continue digging in towards the slope until the trench is one ski pole length long. Be careful not to disturb the area surrounding the shear block. Make sure the walls are even and vertical before cutting the uphill wall.

- Be very careful! Use a snow saw, a piece of string, a ski, or anything that will cut through the snow, and cut the uphill wall.

- Carefully ski to the side of the test site and circle around approaching the upper cut of the shear block diagonally from above. Once skis are perpendicular to the cut on the uphill side of it, gently move to the downhill side of the cut. The steps in figure A-10 should be done in the order stated. Give a reason next to each step explaining the condition of the snowpack if it happens to fail on that particular test. What you will be looking for is at which step of the test the shear block slides.

STEP	PROCEDURE	RESULT
1	Snowpack fails while excavating the test site.	Extremely Unstable.
2	Snowpack fails while approaching the test site.	Extremely Unstable.
3	Snowpack fails while standing on the shear block.	Extremely Unstable.
4	Snowpack fails while flexing your knees.	Unstable
5	Snowpack fails with one jump while wearing skis	Unstable
6	Snowpack fails after repeated hard jumps while wearing skis.	Relatively stable
7	Snowpack doesn't fail after repeated hard jumps while wearing skis.	

Figure A-10 Steps of the Rutschblock (Shear Block) Test

b. Snowpit Analysis. In the snowpit analysis, we deal only with the basic observations. These observations are obtained by digging snowpits and looking for stability and instability keys.

(1) Snowpits should be dug as close as possible to the starting area of the suspected avalanche. They should be at least at a similar elevation, slope observation, and slope angle. Danger is involved. Safety precautions must be taken. The personnel involved can be caught should a slide occur. A good belay should be set up.

(2) Snowpits are relatively easy to dig on steep slopes because depositing of snow is easy. A pit deeper than 2 meters is usually not needed unless no idea of the conditions in the area exists.

(3) The face of the pit should be shaded if possible to prevent melting of the snowpit wall.

(4) To analyze the snowpit-

- Determine the snow depth. If the pit does not reach to the ground, use a probe. Probe several times with a steady push and feel for differences in the snowpack.

- Smooth the pit wall with the tip of the shovel to bring out variations in the layers.

- Look for obvious weak layers such as surface hoar, advanced Temperature Gradient (Depth hoar), ice crusts, graupnal crystals, and cohesionless grains.

- Run the tip of a straight edge down the pit wall with equal force and feel the different layers.

- Look at the crystals of suspected layers.

- Place thermometers at different layers of the pit wall. Look for a Temperature Gradient in the snowpack as well as in the layers. A Temperature Gradient of 10 degrees per meter is an avalanche danger. Layer temperatures will indicate rate and type of metamorphism.

- Use the hand hardness test (see fig. A-11) to determine snow layer density. This test will indicate the stability of the snowpack. These results are further interpreted to establish the potential risk of avalanche danger in the snowpack. A considerable amount of experience and training is necessary to interpret this data. A potentially unstable slab configuration would have a cohesive 1finger hard layer resting on top of a less cohesive first layer. This, in turn, could be underlain by a harder bed (sliding) surface with hardness ranging anywhere from 4 finger to knife. Remember that the strength of a layer is determined by how well the grains within it are bonded. While strong layers are often hard and weak layers soft, this is not always the case. New powder snow can form cohesive slabs despite the fact that it might only be fist hard. The degree of hardness is shown in figure A-11.

F	(Fist)	Determines a very soft layer.
4F	(Four Fingers)	Determines a soft layer.
1F	(One Finger)	Determines a medium layer.
P	(Pencil)	Determines a hard layer.
K	(Knife)	Determines a very hard layer.

Figure A-11 Hand Hardness Test

c. Shovel / Ski Shear Test. The shovel/ski shear test indicates if the previous assumptions in estimating weak layers were correct.

- Cut a wedge column 20 inches wide at the pit wall face and 12 inches wide from front to rear.

- Work the shovel / ski behind the column and exert pressure as you go down looking for separation of the layers. These separations indicate the typical weak layers.

d. Hasty Pit Tests. Digging hasty pits can be quick and effective. They are encouraged during rapid movement in questionable conditions. The purpose of hasty pits is to identify weak layers in the snowpack and determine their strength, and to find the depth of potential slabs that may or may not exist. The methods of analysis with a hasty pit are similar to those of snowpits but modifications are made in testing procedures to obtain essential data for evaluation in less time. Observe obvious weak layers while digging. Conduct the hand hardness test and the shovel shear test. The depth of the hasty pit will be determined by the evaluator's suspicions of where a slab condition exists and time available for the evaluation. A general rule of thumb is from 5 feet to the ground surface.

13. Route Selection

In combat operations or training exercises, route selection in mountainous terrain is extremely important. When planners select objectives or when squad leaders conduct a patrol, they must be aware of the dangers. Most avalanches are triggered by their victims. Conditions vary from slope to slope. Slopes that are safe at 1,000 feet may be death traps by 1,200 feet, or one side of a gully may be stable and the other side unstable.

a. Avalanche Maps. Avalanche maps are available in contingency areas, e.g., Europe. Maps will need to be developed for use in training areas. They identify danger areas where avalanches are expected to occur annually.

b. Weather / Environmental Factors. Past and present weather conditions should be analyzed before operating in avalanche areas. A thorough understanding of the weather and environmental conditions as indicated in paragraphs 1 through 9 is necessary.

14. Crossing Avalanche-Prone Slopes

Operational necessity may require the crossing of a suspected slope. This should be attempted only after all alternatives have been exhausted.

a. Procedure Before Crossing

- Equip Marines with special equipment like snow shovels, probe poles, and transceivers, if available.

- Keep the pack on and secured.

- Secure ECWCS hood and all open pockets

b. When Crossing Avalanche Zones. Post a spotter who observes as each man crosses. Cross one Marine at a time using a colorful avalanche cord of approximately 100 feet, if possible. If an individual is buried, the spotter will have noticed him somewhere in the path and the avalanche cord will float to the top. Travel quickly from one safe location to another to minimize the time exposed to danger. Try to use the same tracks to minimize disturbance to the snow and the time exposed to the hazard.

c. Avalanche Transceivers. The biggest innovation in avalanche rescue in recent years has been the introduction of electronics rescue transceivers. They transmit electromagnetic signals that can be picked up by another transceiver in a receiver mode. They are a fast, effective method of locating avalanche victims. Their use requires practice.

- Remember that carrying the transceiver does not give a license to take additional risks.

- There will be a constant need for fresh alkaline batteries.

- Be sure that both the avalanche rescue party and the victim use compatible transceivers (frequencies).

- Set the transceiver on and test all transceivers for transmit and reception before the ski march begins. It will be difficult if not impossible to do this once caught in an avalanche.

- Carry the transmitting unit inside the clothing suspended by straps. A high-speed avalanche can remove a surprising amount of clothing.

- If possible, cross all avalanche paths one man at a time. There will be competing signals in a multi-burial avalanche, which will make location of a single victim more difficult.

- Do not-

    - Store the transceiver without batteries because of moisture and battery leakage.

- Place the transceiver equipment into the pack. The pack will probably become lost in an avalanche.

- Wear a quartz watch; it will cause a disturbance in the earphone.

- Remove the transceiver until the end of the ski march.

(1) Frequencies. Marine units operate and conduct training both in CONUS and with allies overseas. Two different frequencies exist. In the United States they operate off of 2.275 KHz, and in European countries they use 457 KHz. Be sure that compatible frequencies are used by all members of your unit. When operating overseas, the frequency selected should be compatible with the standard host nation search and rescue frequencies.

(2) Bracketing. The depth of burial does not affect the signal range but it does affect the volume (See fig. A-12).

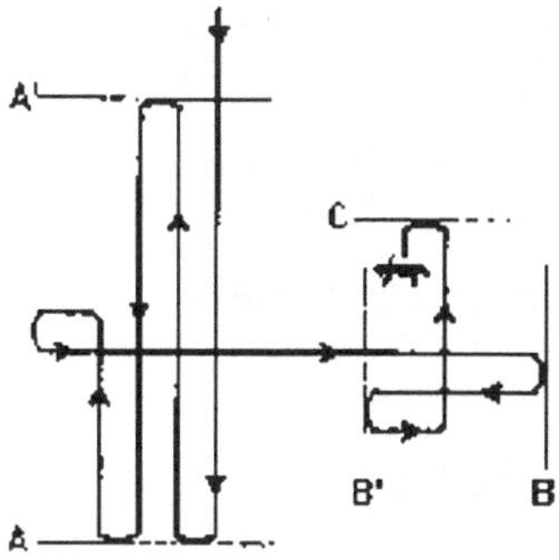

Figure A-12 Transceiver Bracketing

- Be sure that transceivers are switched to RECEIVE.

- Deploy a line of searchers at a maximum of 30 meters apart at the last point where the victims were last seen and move down the slope.

- Volume control should be all the way up until the first signal is received.

- Move in unison on line stopping every 10 paces, rotating the transceivers to check for signal.

- When the first signal is heard, everyone is informed but the line should not break. Bracket to within 3 to 6 feet of the victim. Searchers probe to find the victim to avoid having to dig up too much snow and so time is not lost.

14. Avalanche Search and Rescue

Statistics show that after ½ hour of burial, the chances of survival are about 50 percent, after an hour, 20 percent. Speed is essential for recovering a live victim! Cold and suffocation are the main causes of death. Always carry a shovel! In training, a search should not be called off for at least 24 hours. In combat, the tactical situation will dictate. Air pockets have kept victims alive for several days. One victim survived an avalanche in a structure for 7 days but lost both feet to frostbite. No one has been found alive deeper than 7 feet.

a. <u>Most likely spots to find an avalanche victim</u>. The search begins at the last location where the victim was seen. Look for clues such as skis, clothing, or avalanche cord. Concentrate the search at the outside of bends where debris accumulates. Look for victims on the uphill side of obstacles, such as trees and boulders where debris builds up. In the runout zone, debris may be very large and hard to search in.

b. <u>Equipment</u>. A unit should be properly equipped with the following equipment. If not available, quick replacements may be improvised.

- Enough probes for 2 probe lines.
- Markers. (Chemlites)
- Snow shovels.
- Splints
- Rewarming Bags.
- Surveyor Tape.
- Flashlights.
- Litters or Backboards.
- Resuscitators and oxygen.
- Transceivers.

15. Types of Search

The type of search conducted may depend on available manpower and time.

a. <u>Hasty Search</u>. The hasty search is by far the most important search for military operations. Speed is essential in the initial search. An immediate decision must be made as to whether additional assistance is required.

Generally, a military unit will have the communications necessary to send for assistance. This type of search should continue until the company commander makes a decision to call it off.

(1) One or two Marines leaving to get help can greatly degrade the search effort of a small group. In larger units, teams can split up to look for clues and to probe likely spots. Probing is best accomplished using a standard probe pole. However, ski poles or sticks may be used. Be sure that avalanche guards are posted! *Always be aware of secondary avalanches!*

(2) Conduct the hasty search using the course probe as shown in figure A-13. The course probe search sacrifices some thoroughness in exchange for speed. Two squads are designated to the probe line. Markers and digging teams are designated from the remaining squad. The platoon sergeant is in charge.

Figure A-13 Hasty Search

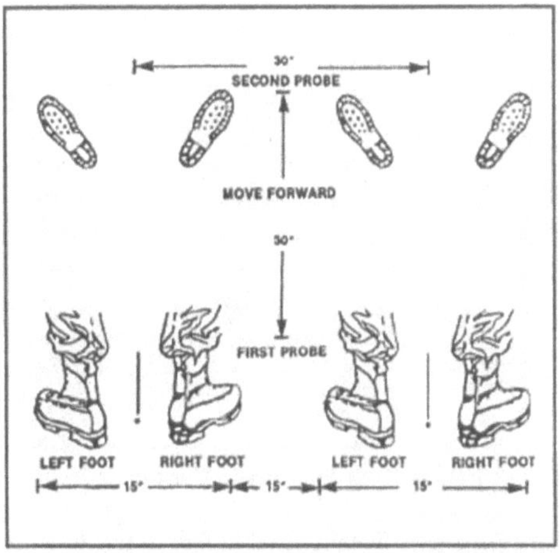

-With two squads on line at close interval, each man places the probe between his feet.

-The platoon sergeant gives the command Down Probe. Probes are pushed down through each layer of snow and ice being careful not to pierce a victim. Every strike (Potential Victim) must be marked and uncovered.

-The platoon sergeant then gives the command Up Probe. Probes are withdrawn.

-The platoon sergeant then gives the command Step. Each Marine takes a 30-inch step, being careful not to step on the hole made by his probe. The probe hole will allow a victim's scent to rise through the snow more quickly for the search dogs, if they are available.

-Repeat this process until the victim is found.

-If a strike (find) is made, the prober signals the markers and a flag is placed at the spot indicated by the prober. The line never stops.

-Shovelers attempt to uncover a strike, digging all the way to the ground.

-Strikes are uncovered in turn. However, the platoon sergeant may determine exceptions.

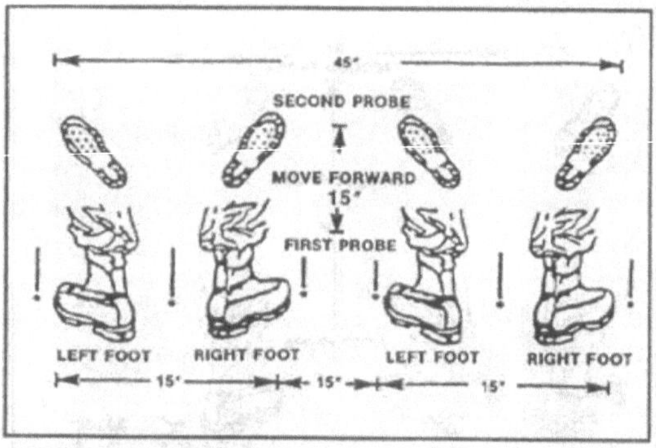

Figure A-14 Fine Probe Search

b. Fine Probe Search. A fine probe takes four to five times longer than the course probe. Because it takes longer, chances of the victim being recovered alive are greatly reduced. Execution is the same as the course probe except a 15-inch step is taken and probing is done over the left, middle, and right (See fig A-14).

c. Specialized Searches. Conducting specialized searches depends on the availability of specialized equipment that Marines may have available to them; e.g., transceivers and infrared devices and trained avalanche dogs.

Host nations sometimes provide avalanche dogs. Marines may be trained to use transceivers.

d. Unit Avalanche Search. Figure A-15 show how a unit would task-organize its members to conduct an avalanche search if a portion of the unit has been caught in an avalanche. The remainder of that unit or a unit that is assigned to conduct the rescue should deploy its members as shown in figure A-15.

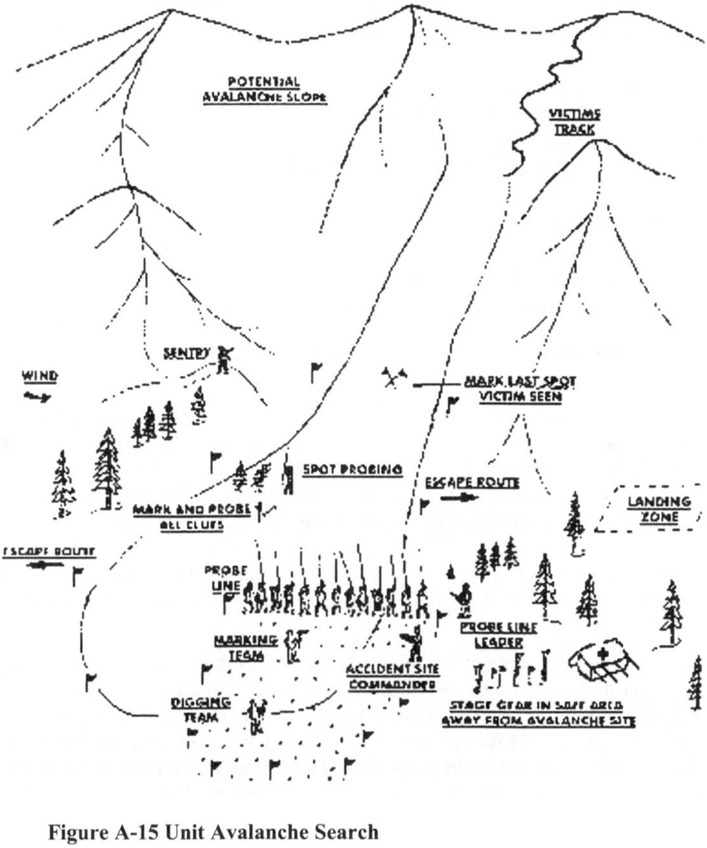

Figure A-15 Unit Avalanche Search

APPENDIX B: FIELD WORKS AND CAMOUFLAGE

A. Field preparations in cold weather conditions are basically the same as in temperate climates, but with certain variations. The cold weather conditions make it necessary to improvise. The most important factors to consider are: frozen ground, snow, cold, darkness. The above factors lead to the following:

- Reduced work capacity.
- The need for heating at the field positions.
- Digging made difficult because of frozen ground.
- Increased need for explosives in building positions.
- Reduced mobility.
- The necessity of clearing snow.
- Lakes and rivers no longer hinder the enemy's advance.
- Trails are difficult to conceal.
- It is more difficult to make use of natural camouflage.

The purpose of this publication is to give some guidance in how best to manage in cold weather conditions with the means available.

B. Field Fortifications. When field fortifications are to be constructed it is difficult to decide whether it is best to dig into the frozen ground, or to build above ground. The decisive factor will normally be the amount of explosives that have been allocated for digging and usually that have been allocated for digging and usually a compromise between the two options will have been reached.

Weapon siting, in particular the elevation and depression requirements of the various weapons, will be another important factor to consider when deciding whether to dig in or build up.

Firing positions should not be built in such a manner that riflemen have to be in the prone position, because the risk of frostbite is reduced when kneeling or standing. In the prone position there is little possibility of moving the body, and a large part of the body touches the cold surface beneath. The firing position should be built in such a manner that the soldier could stand up, and preferably move about within it, or in communication trenches. Snow reduces the preparation of small arms and shell fragments. (Figure B-1)

MATERIAL	MINIMUM THICKNESS

Newly fallen snow	400 cm
Packed Snow	200 cm
Frozen snow / water mixture (snowcrete)	150 cm
Ice crete	50 cm

Figure B-1 Bullet Penetration Table

1. Machine gun fire from 300m range has the same effect as the rifle fire. When the range is shorter, the penetration of machine guns is increased by about 50 percent. If a firing position is to be splinter proof, heavier material is required for revetting the firing position and shelter.

2. Ice crete can be used for secondary fortification of a position. Sand and water are mixed and pressed within forms or containers preferably reinforced with twigs, wire, and similar things. Water can be poured over the frontal cover to make it stronger.

3. Standard unit and firing positions are to be built, as in summertime, but with modifications as demanded by cold weather conditions. The amount of snow, ground conditions, and available construction equipment decide whether positions should be above ground, partially dug in, or completely dug in.

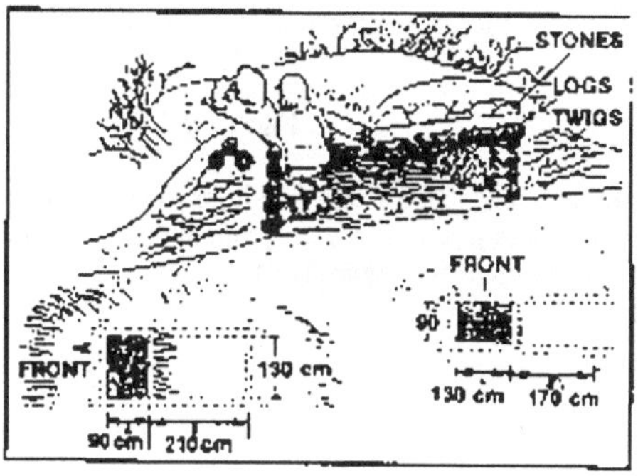

Figure B-2

4. The following rules are to be observed when building a position in snow (above ground, or partially above ground):

- The position should be sited in a place where the terrain can be used to the greatest advantage and can provide geed frontal cover.

The position should be as low as possible on the ground.
Natural camouflage should be used.
The snow in front of the position should not be removed.

- The position can be strengthened with timber, sandbags, and ammunition crates filled with sand or ice crete.
- The height of the position must correspond to the amount of snow and terrain, but the height inside the shelter must not be less than 70 cm.

- The snow around the position should be packed as hard as possible. If more snow is pilled on and the packing repeated the snow would gradually freeze and become quite compact.
- Sticks, twigs, earth, etc., should be embedded towards the front of the snow defense, to stop projectiles and shell fragments as far forward as possible.

5. If revetting material is in short supply, it is best to build an improvised firing position, and let shelter wait. Priority must be given to building a strong frontal cover (refer to figure B-3).

Figure B-3

6. When the position is to be completely dug-in, the usual construction practice is followed. In such cases when the frozen ground has been burst through, it is easy to improvise a good position (See figure B-4).

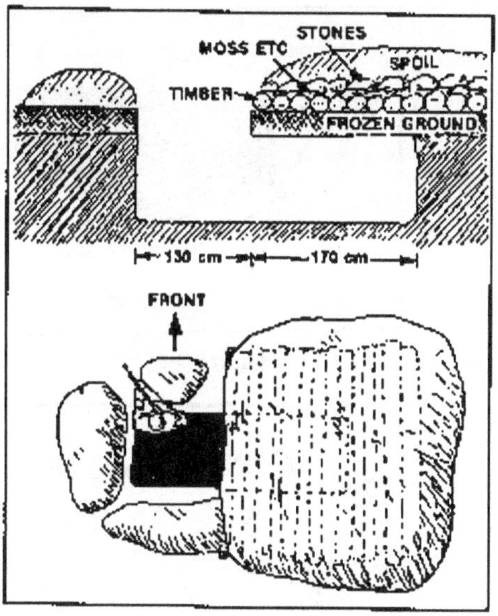

Figure B-4

- The sides of the trench become excellent walls.
- If a larger position is wanted, a shelter can be built below the layer of frozen ground.
- Vertical supports are erected beneath to strengthen the overhead cover.
- The overhead cover is reinforced by placing timber on it and by piling the dugout earth on top.
- It is important to camouflage all dugout earth with snow.

Marsh areas are unsuitable for dug-in positions, because the seepage of water is too great and the layer of frozen ground too thin.

7. Large shelters are built in the same manner as in summertime. The frozen ground is dynamited, or broken through with pioneer platoon drilling equipment.

8. Small-improvised shelters and covers can be built by modifying the 4-man shelter, which is built as a firing position but with a cover over the entire position. The height can vary according to the conditions (refer to figure B-5).

   Small-improvised shelters can also be built by utilizing the terrain in such a manner that a complete built-up shelter becomes unnecessary. A splinter-proof cover will in certain cases give full protection against indirect fire (refer to figure B-6).

   In extreme cases a shallow shelter can be made for an individual lying down or sitting with splinter-proof overhead cover (refer to figure B-7).

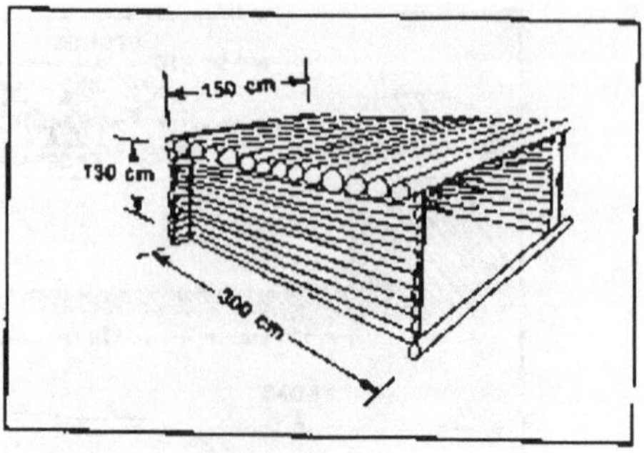

Figure B-5

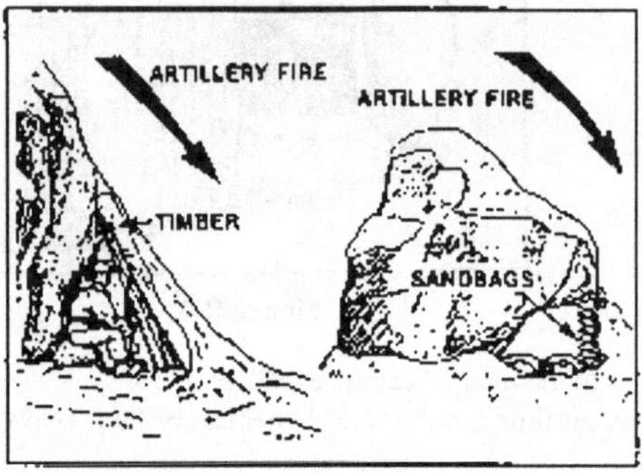

Figure B-6

Figure B-7: Position for Prone Rifleman with Splinter-proof Overhead Cover.

9. The improvised position is to meet the standard requirements as regards construction and protection. The need for thorough control and inspection becomes greater when improvising positions.

C. Communication Trenches. Communications trenches in snow are built after the same principles as in earth. Dugout snow is used to strengthen the side facing the enemy. Connecting trenches in snow are primarily used to provide concealment; it takes a lot of snow to provide protection against enemy fire. The bottom of the communications trench should be covered with snow to make observation from the air difficult. The snow dugout near the firing position should be used to increase the thickness of the frontal cover.

D. Explosives. During cold weather, explosives have to be used in new ways to meet operational requirements. Frozen ground has to be blasted open in the building of positions, and gaps may have to be blasted in the ice on rivers or lakes to canalize enemy movement and reduce the need for own forces. Explosives and fuses are exposed to cold and dampness, and it is necessary to know how this affects usage and handling. Explosives that can stand the cold are called frost-free.

1. Military explosives are frost-free. TNT will slowly be ruined by dampness but over-all is the best and most versatile military explosive. C-4 becomes hard and difficult to work with in the cold. Detonation cord becomes stiff and near impossible to work with in the cold as well. Try not to bend the cord as it may cause breakage in the PETN and misfires may occur. Dynamite is water-resistant and remains virtually unchanged in the cold but does not have many uses. For these reasons it is important to prep all charges in a warm environment if possible.

2. Initiation. Time fuse remains unchanged by the cold but is affected by altitude due to the thin air. Test all time fuse prior to use. The M-60 fuse igniter is not waterproof and can become unusable due to dampness. Always dual prime all shots to ensure detonation.

3. Extreme cold temperatures due to increased static electricity in the air can affect electronic detonators, but over-all remain unchanged.

4. Blasting through Frozen Ground. The aim is to break through the layer of frozen ground. The most difficult part is to make the initial hole through it so that the charge can be inserted for further blasting. This can be done in the following ways:

Drill a hole with a pneumatic or hand drill from the pioneer kit.
Blast a hole for the main charge.

5. Placing of Charges in Frozen Ground. The hole is made with pneumatic or hand drill. The depth of the drill hole should be 9/10ths of the depth of the frozen ground (refer to Figure B-8). The interval between the holes and between rows of holes should not exceed the depth of the frozen ground. The charges should fill up 2/3rds of the depth of the holes.

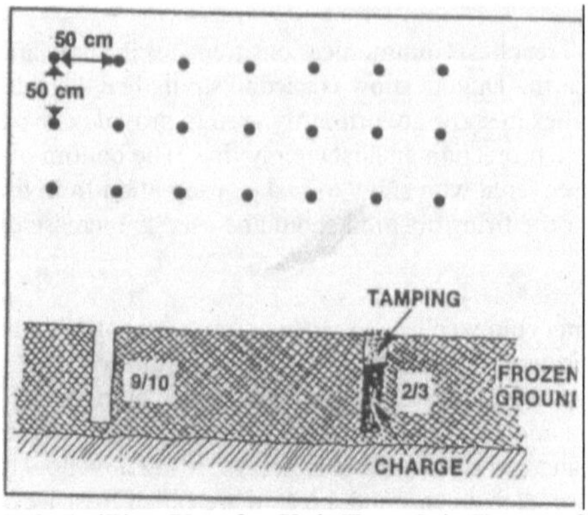

Figure B-8: Drilling Plan for Unit Trench in Frozen Ground.

When improvising a position, a smaller trench is sufficient; the shelter can then be built underneath the layer of frozen ground.

6. Blasting of Holes in Frozen Ground. Holes are blasted in the frozen ground with the following means:

Shaped charges.
Pressure or tamped charges.

A tamped/pressure charge can be used when a layer of frozen ground is shallow. (See fig. D-9) The use of a 15lb shaped charge is also effective and will make a suitable hole for placing a charge.

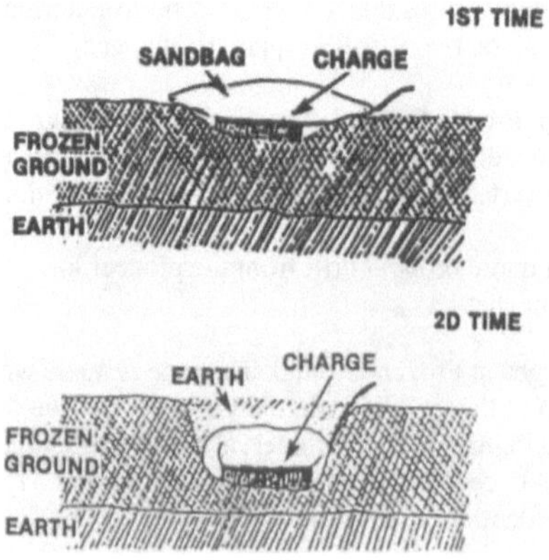

Figure B-9: Use of Pressure Charge

7. The blasting of frozen ground will demand a number of charges. With an interval between them equal to the depth of the frozen ground. If the frozen ground is 50 cm deep, fourteen holes are required, in two rows, with 50 cm intervals between the holes and 50 cm between the rows. If the frozen layer is thick, it might be necessary to repeat the blasting. The shaped charge will often pierce the layer of frozen ground. To achieve the full effect of the blast, the hole must be prepared as in figure B-10 before the second charge.

8. If a unit trench is to be made by blasting craters, one to three holes are needed. The holes are blasted through the frozen ground; then the hollow blasted beneath the frozen ground is filled in to make a suitable chamber for the main charge (refer to figure B-11). If necessary, the hole can be enlarged by the use of dynamite.

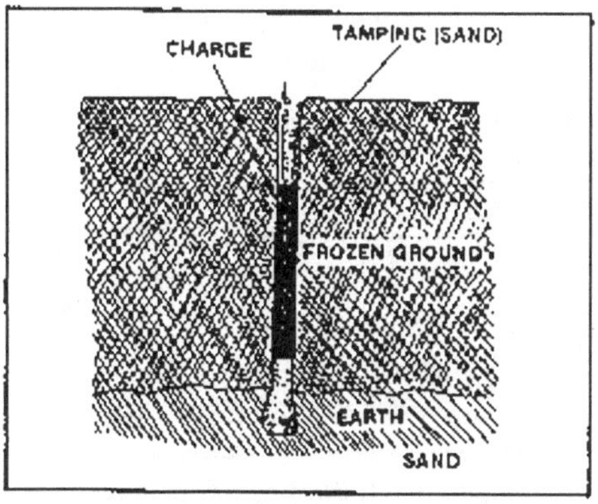

Figure B-10. Preparation of Hole Made by Shaped Charge

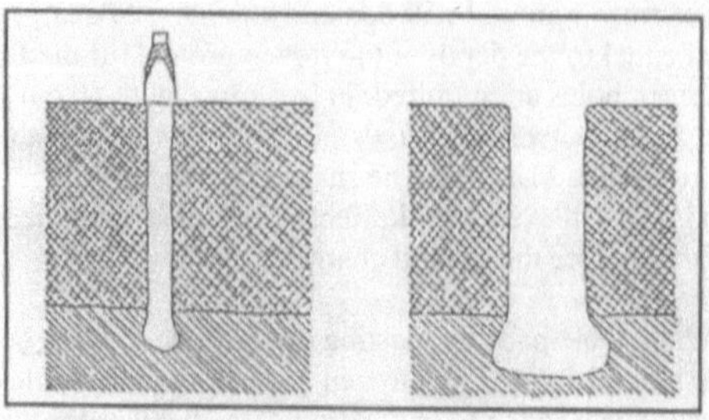

Figure B-11. Shaped Charge and Main Charge Craters

E. Ice Demolition. The purpose of blasting away ice is most often to open up a clear water gap to block the enemy. Detailed reconnaissance is necessary to get the gap opened where the current is strongest. An ice drill is needed, an axe, a tape measure, a sounded line, and the blasting equipment. The thickness of the ice and the depth of the water are measured by drilling a series of sounding holes (intervals of 50-100m). When the location of the gap has been decided, the centerline should be marked with a ski trail, twigs, or in some other manner.

1. A charge detonated underneath the ice throws a circular mound of water up, breaking the ice. If slush remains in the gap, it will freeze quickly in frigid temperatures.

2. Estimation and Placing of Charge. A gap is blasted in the ice by placing a series of charges underneath it (refer to figure B-12).

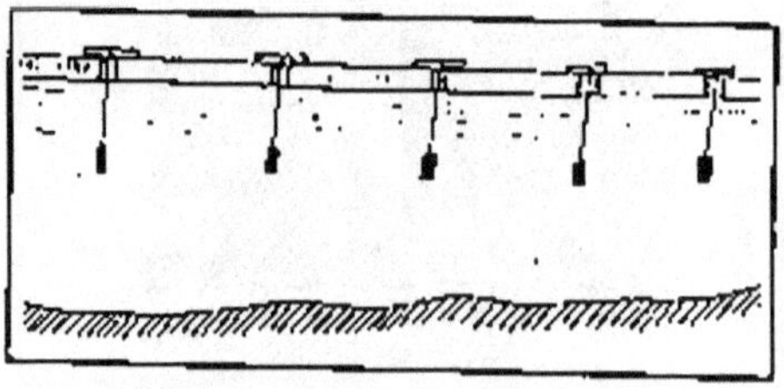

Figure B-12: Blasting a Gap in Ice

3. If the depth of the water is greater than 2.5m, the charge should be placed about 1.25 m beneath the ice. If the depth of the water is less than 2.5 m the charge should be placed at half the depth to get optimum results. The size of the charge should be as in figure B-13.

THICKNESS OF ICE	WIDTH OF THE GAP	WEIGHT OF CHARGE	INTERVALS
up to 40 cm	5 m	1 kg	4 m
up to 40 cm	8 m	2 kg	5 m
up to 40 cm	8 m	3 kg	8 m
40 – 60 cm	8 m	4 kg	8 m
60 – 100 cm	8 – 10 m	5 kg	8 m

Figure B-13: Table of Charges-Depth more than 2.5 m

If the depth of water is less than 2.5 m, the hole in the ice made by any given charge becomes smaller. Therefore, the interval between the charges is reduced as shown in figure B-14

DEPTH OF WATER	INTERVAL BETWEEN CHARGES		
m	3 kg charge	4 kg charge	5 kg charg
2.0 m	5 m	7 m	9 m
1.5 m	4 m	6 m	8 m
1.0 m	4 m	5 m	8 m
0.5 m	3 m	4 m	5 m

Figure D-14. Table of Charges—Depth of Water Less Than 2.5 m.

4. TNT. C-4, and dynamite will all work as the base charge for ice breaching. Dynamite needs to be water proofed prior to use. The charges are suspended as shown in figure D-15.

5. To make a hole in the ice for the main charge, the following can be used:
   - Motor-powered ice drill.
   - Ice auger (hand for tactical purposes).
   - Explosives.
   - Axe, Pick, and crowbar.

If explosives are to be used, a hole should first be made in the ice, and the charge should be 100 grams per 10 cm of ice.

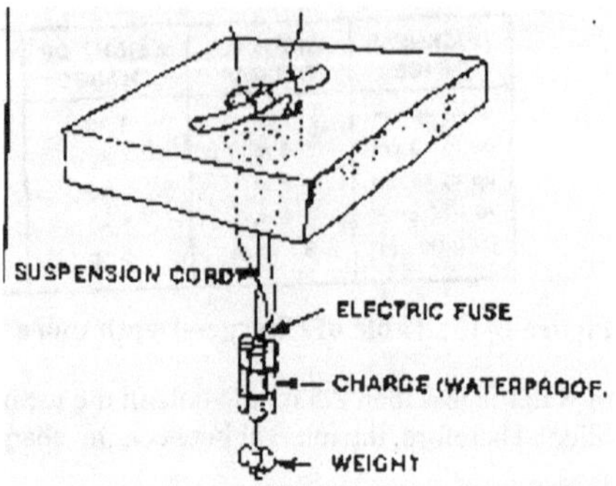

Figure B-15. Example of Suspended Charge for Demolishing Ice

6. Blasting / Breaching the Ice. The ice breaching party is organized into three teams:

   a. Reconnaissance Team. Selects appropriate site and determine the demolition requirements and report this information to the demolition team. Prepares holes as required.

   b. Demolition Team. Prepares charges in a warm, concealed position if possible. The team brings the charges on to the ice and places them properly in the holes.

   c. Initiator. Checks charge and initiate the shot. Upon detonation the team will inspect holes and check for unexploded ordnance.

7. Tactical Uses. The purpose opening gaps in the ice is to protect units against frontal and flanking attacks as well as impeding enemy travel. Breaching ice can also be used as an ambush technique against enemy patrols.

8. Water obstacles will begin to refreeze in frigid temperatures. Friendly forces must be prepared to conduct ice maintenance as needed. Covering gaps/holes with tarp or plastic sheeting can do this.

F. Artificial Obstacles. Obstacles built in a warm climate are the constructed the same with certain modifications due to the snow and frozen ground.

1. Obstacles in Frozen Ground. When timber, steel pickets, etc. cannot be driven into the ground, they can be lashed together with vertical and horizontal wraps as shown in figure D-16. When placing the obstacle on ice, construct footings for the legs, fill with snow and water to cement them in.

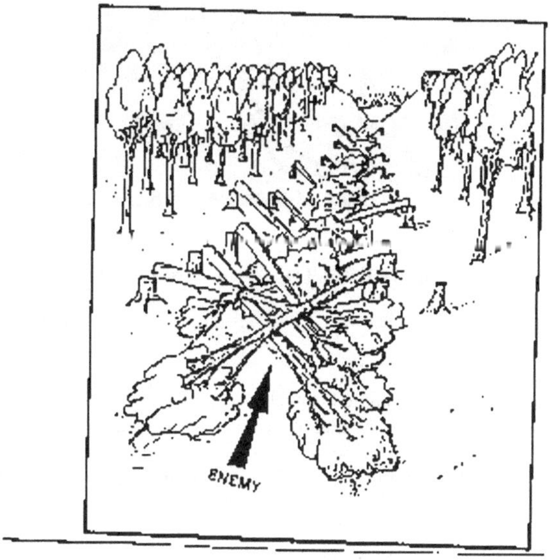

Figure B-16. Abatis.

2. Obstacles on Snow. Construct a horizontal beam and attach it to the butt ends of the obstacle or attach barbed wire or concertina to act as a floatation device. For added stability, anchor the butt ends.

3. Concertina laid on the snow (single/double/triple strand) is an easy and simple obstacle to employ against enemy skiers (refer to figure B-17).

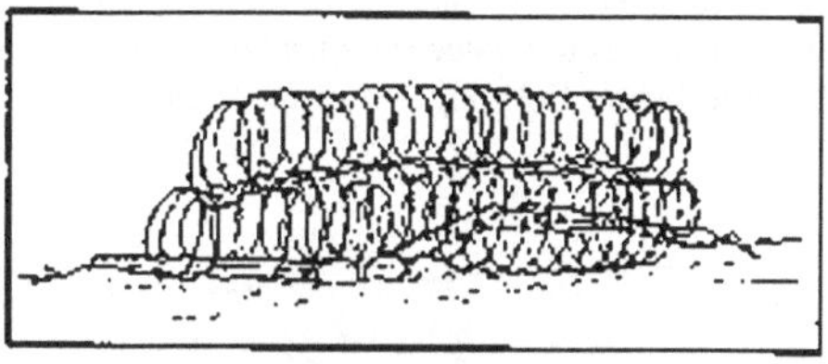

Figure B-17: Concertina on Snow

4. Using trip wires and AP mines on known enemy ski trails is an easy and effective obstacle to employ (refer to figure B-18).

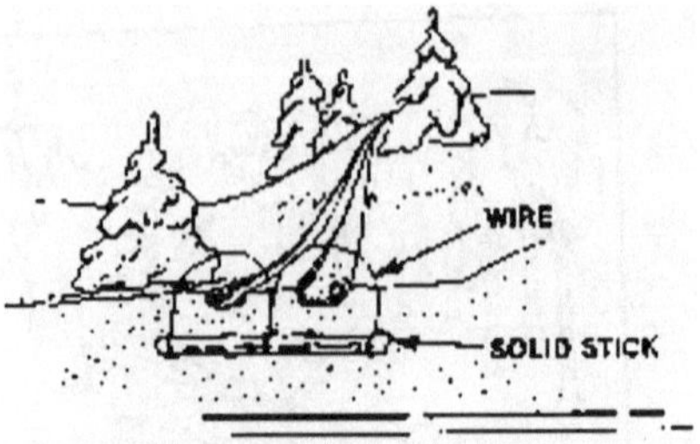

Figure B-18: Trip Wire on Ski Trail

5. Reinforce all obstacles with AP mines and trip flares to add to their effectiveness.

6. Anti-Armor, Anti-vehicle Obstacles. With the winter come added problems for vehicle travel. Snow and ice is a major problem in itself. When conducting reconnaissance, it is important to find lines of natural obstacles that require little improvement. The effort must be concentrated at the parts where the line is weakest.

7. Rivers, lakes, snow covered slopes, snow and ice, forest, and avalanche zones are just some of the natural obstacles that a winter environment brings that can impede vehicle/armor travel.

   When selecting a position, the location of natural obstacles must be carefully considered. Frequently there is more than one natural obstacle in the same place. The total effect must be carefully weighed.

8. Natural obstacles are improved when necessary:

Explosives can be used to break open waterways, create craters, lanes, and tank ditches. Obstacles can be built on slopes to reduce the speed of advance; examples of these are the timber barricade (refer to Figure B-19), the snow barricade, and felled trees such as an abatis.

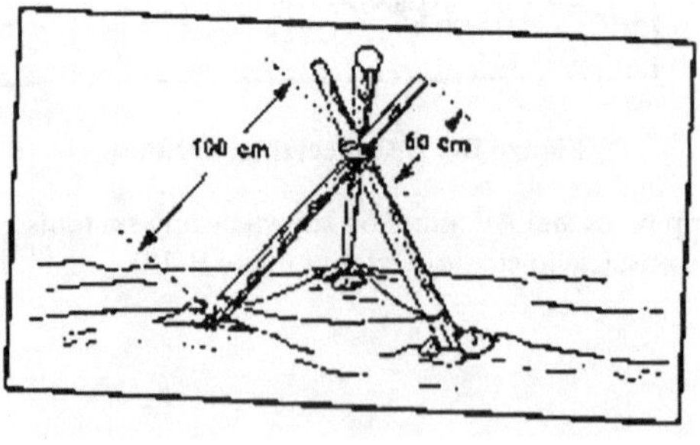

Figure B-19 Timber Obstacle

G. Mines are presently the most important passive means available for combat operations. The preparation, placing, and concealing of mines must receive particular attention in wintertime. The fuses and mechanical parts of the mine must be free of storage grease, for the grease can cause problems in very cold weather. The placing of mines in plastic bag waterproofs the mine to a degree but condensation in the bag during times of fluctuating temps can cause problems. A dark mine absorbs heat in the day and freezes at night. It is difficult to conceal mines for prolonged periods of time. The painting of the mines white could help

1. Anti-personnel Mines. It is easy to conceal anti-personnel mines in the snow. The use of trip wires and anti-handling devices can be used but with extreme caution due to the ever-changing climate and problems that may arise with it. M605 fuses for the M16A2 should be cleaned with diesel to lubricate and preserve the fuse. Replace both safeties with safety pins to simplify the employing and removal process.

   a. All natural obstacles should be reinforced with AP mines. It is important to note that snow will soak up a great deal of the blast. All mines should be placed on a platform when employing to increase their effectiveness.

   b. A hole can be dug to the side of a ski trail. Place the mine on a platform. Slide a board under the ski track horizontally; place a stick or small piece of wood vertically on the other side of the ski track to support the horizontal board. Ensure that the prongs are in positive contact with the horizontal board, then arm and gently bury the mine (refer to figure B-20.).

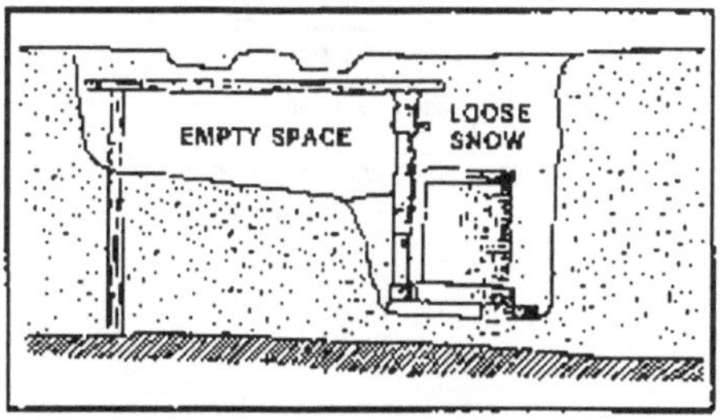

Figure B-20: Mine under Ski Trail

2. It is difficult to camouflage an anti-tank minefield in a snow-covered environment. As a rule the best place to employ an anti-tank minefield is in a well-tracked area.

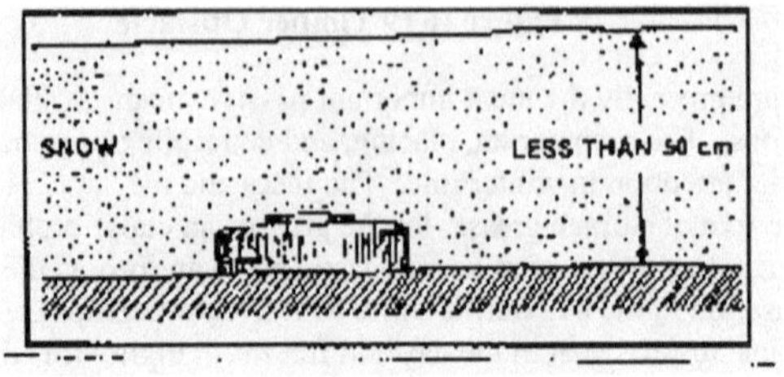

Figure B-21: Positioning the Anti-tank mine M-15

If the snow depth is 50 cm or less, placing the M-15 directly on the ground will not reduce its effectiveness. If the snow is deeper, packing the snow and using a wooden platform to rest the mine on will increase the effectiveness. A common practice is to use two mines stacking one on top of the other. The bottom is turned upside down and armed while the top is placed right side up and armed. This does two things, one it gives the top mine a platform, two the bottom mine is protected from the elements. Water seeping into the mechanical features of the bottom mine is reduced significantly. Refer to Figures B-22 and B-23.

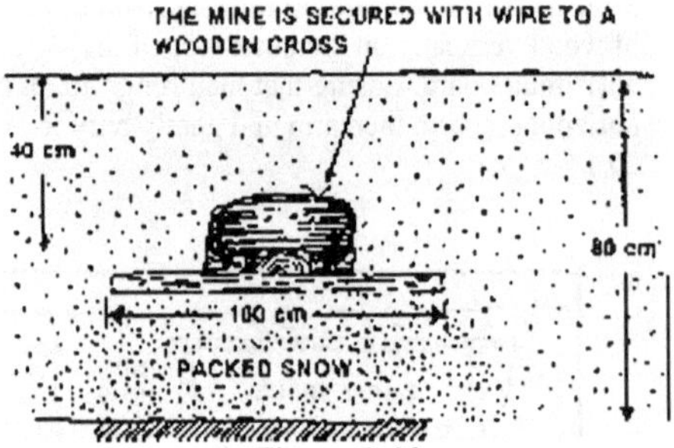

Figure B-22

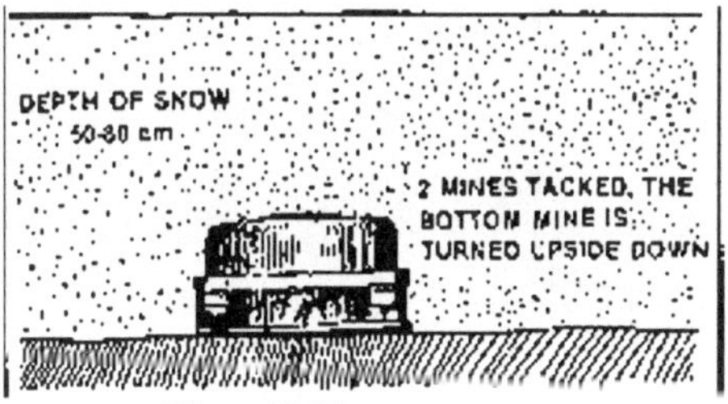

Figure B-23

Hasty surface laid minefields are still effective in this environment. They are quick, not gear or personnel intensive, and is easily employed and removed. When covered by fire they reduce the enemies' mobility heavily.

APPENDIX B: FIELD WORKS AND CAMOUFLAGE

A. Field preparations in cold weather conditions are basically the same as in temperate climates, but with certain variations. The cold weather conditions make it necessary to improvise. The most important factors to consider are: frozen ground, snow, cold, darkness. The above factors lead to the following:

- Reduced work capacity.
- The need for heating at the field positions.
- Digging made difficult because of frozen ground.
- Increased need for explosives in building positions.
- Reduced mobility.
- The necessity of clearing snow.
- Lakes and rivers no longer hinder the enemy's advance.
- Trails are difficult to conceal.
- It is more difficult to make use of natural camouflage.

The purpose of this publication is to give some guidance in how best to manage in cold weather conditions with the means available.

B. Field Fortifications. When field fortifications are to be constructed it is difficult to decide whether it is best to dig into the frozen ground, or to build above ground. The decisive factor will normally be the amount of explosives that have been allocated for digging and usually that have been allocated for digging and usually a compromise between the two options will have been reached.

Weapon siting, in particular the elevation and depression requirements of the various weapons, will be another important factor to consider when deciding whether to dig in or build up.

Firing positions should not be built in such a manner that riflemen have to be in the prone position, because the risk of frostbite is reduced when kneeling or standing. In the prone position there is little possibility of moving the body, and a large part of the body touches the cold surface beneath. The firing position should be built in such a manner that the soldier could stand up, and preferably move about within it, or in communication trenches. Snow reduces the preparation of small arms and shell fragments. (Figure B-1)

MATERIAL	MINIMUM THICKNESS

Newly fallen snow	400 cm
Packed Snow	200 cm
Frozen snow / water mixture (snowcrete)	150 cm
Ice crete	50 cm

Figure B-1 Bullet Penetration Table

1. Machine gun fire from 300m range has the same effect as the rifle fire. When the range is shorter, the penetration of machine guns is increased by about 50 percent. If a firing position is to be splinter proof, heavier material is required for revetting the firing position and shelter.

2. Ice crete can be used for secondary fortification of a position. Sand and water are mixed and pressed within forms or containers preferably reinforced with twigs, wire, and similar things. Water can be poured over the frontal cover to make it stronger.

3. Standard unit and firing positions are to be built, as in summertime, but with modifications as demanded by cold weather conditions. The amount of snow, ground conditions, and available construction equipment decide whether positions should be above ground, partially dug in, or completely dug in.

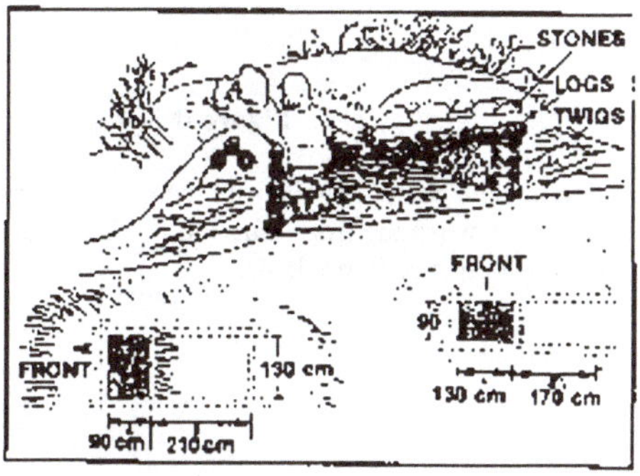

Figure B-2

4. The following rules are to be observed when building a position in snow (above ground, or partially above ground):

- The position should be sited in a place where the terrain can be used to the greatest advantage and can provide geed frontal cover.

The position should be as low as possible on the ground.
Natural camouflage should be used.
The snow in front of the position should not be removed.

- The position can be strengthened with timber, sandbags, and ammunition crates filled with sand or ice crete.
- The height of the position must correspond to the amount of snow and terrain, but the height inside the shelter must not be less than 70 cm.

- The snow around the position should be packed as hard as possible. If more snow is pilled on and the packing repeated the snow would gradually freeze and become quite compact.
- Sticks, twigs, earth, etc., should be embedded towards the front of the snow defense, to stop projectiles and shell fragments as far forward as possible.

5. If revetting material is in short supply, it is best to build an improvised firing position, and let shelter wait. Priority must be given to building a strong frontal cover (refer to figure B-3).

Figure B-3

6. When the position is to be completely dug-in, the usual construction practice is followed. In such cases when the frozen ground has been burst through, it is easy to improvise a good position (See figure B-4).

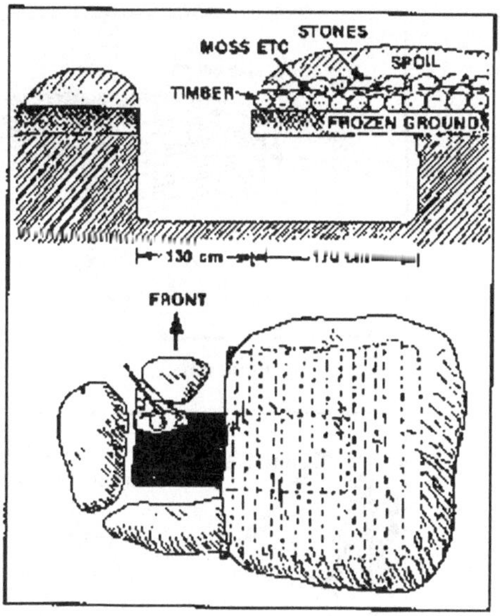

Figure B-4

- The sides of the trench become excellent walls.
- If a larger position is wanted, a shelter can be built below the layer of frozen ground.
- Vertical supports are erected beneath to strengthen the overhead cover.
- The overhead cover is reinforced by placing timber on it and by piling the dugout earth on top.
- It is important to camouflage all dugout earth with snow.

Marsh areas are unsuitable for dug-in positions, because the seepage of water is too great and the layer of frozen ground too thin.

7. Large shelters are built in the same manner as in summertime. The frozen ground is dynamited, or broken through with pioneer platoon drilling equipment.

8. Small-improvised shelters and covers can be built by modifying the 4-man shelter, which is built as a firing position but with a cover over the entire position. The height can vary according to the conditions (refer to figure B-5).

   Small-improvised shelters can also be built by utilizing the terrain in such a manner that a complete built-up shelter becomes unnecessary. A splinter-proof cover will in certain cases give full protection against indirect fire (refer to figure B-6).

   In extreme cases a shallow shelter can be made for an individual lying down or sitting with splinter-proof overhead cover (refer to figure B-7).

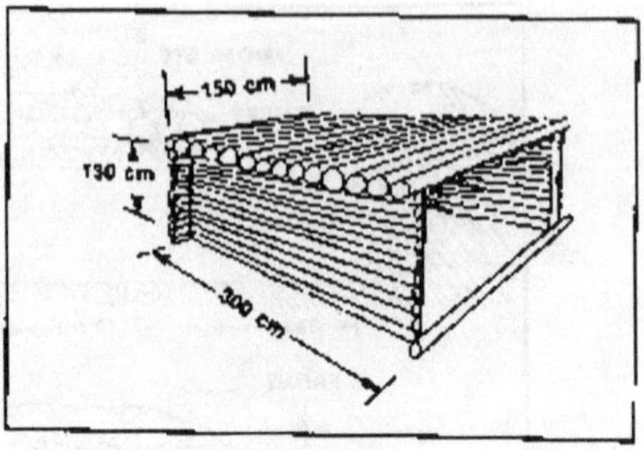

Figure B-5

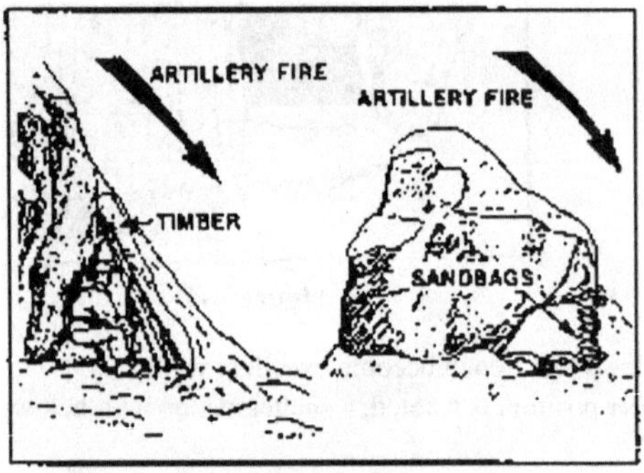

Figure B-6

Figure B-7: Position for Prone Rifleman with Splinter-proof Overhead Cover.

9. The improvised position is to meet the standard requirements as regards construction and protection. The need for thorough control and inspection becomes greater when improvising positions.

C. Communication Trenches. Communications trenches in snow are built after the same principles as in earth. Dugout snow is used to strengthen the side facing the enemy. Connecting trenches in snow are primarily used to provide concealment; it takes a lot of snow to provide protection against enemy fire. The bottom of the communications trench should be covered with snow to make observation from the air difficult. The snow dugout near the firing position should be used to increase the thickness of the frontal cover

D. Explosives. During cold weather, explosives have to be used in new ways to meet operational requirements. Frozen ground has to be blasted open in the building of positions, and gaps may have to be blasted in the ice on rivers or lakes to canalize enemy movement and reduce the need for own forces. Explosives and fuses are exposed to cold and dampness, and it is necessary to know how this affects usage and handling. Explosives that can stand the cold are called frost-free.

1. Military explosives are frost-free. TNT will slowly be ruined by dampness but over-all is the best and most versatile military explosive. C-4 becomes hard and difficult to work with in the cold. Detonation cord becomes stiff and near impossible to work with in the cold as well. Try not to bend the cord as it may cause breakage in the PETN and misfires may occur. Dynamite is water-resistant and remains virtually unchanged in the cold but does not have many uses. For these reasons it is important to prep all charges in a warm environment if possible.

2. Initiation. Time fuse remains unchanged by the cold but is affected by altitude due to the thin air. Test all time fuse prior to use. The M-60 fuse igniter is not waterproof and can become unusable due to dampness. Always dual prime all shots to ensure detonation.

3. Extreme cold temperatures due to increased static electricity in the air can affect electronic detonators, but over-all remain unchanged.

4. Blasting through Frozen Ground. The aim is to break through the layer of frozen ground. The most difficult part is to make the initial hole through it so that the charge can be inserted for further blasting. This can be done in the following ways:

Drill a hole with a pneumatic or hand drill from the pioneer kit.
Blast a hole for the main charge.

5. Placing of Charges in Frozen Ground. The hole is made with pneumatic or hand drill. The depth of the drill hole should be 9/10ths of the depth of the frozen ground (refer to Figure B-8). The interval between the holes and between rows of holes should not exceed the depth of the frozen ground. The charges should fill up 2/3rds of the depth of the holes.

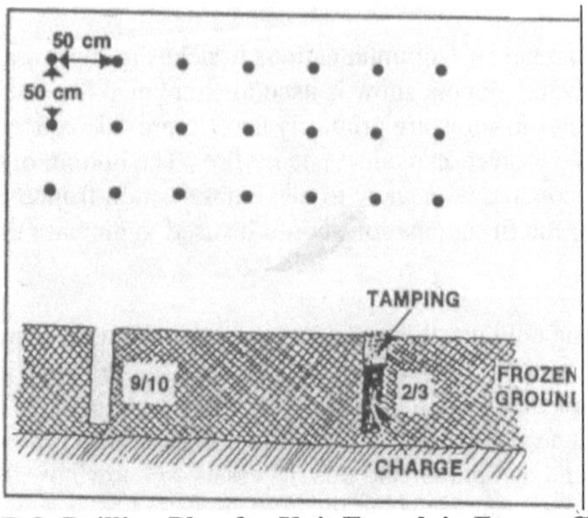

Figure B-8: Drilling Plan for Unit Trench in Frozen Ground.

When improvising a position, a smaller trench is sufficient; the shelter can then be built underneath the layer of frozen ground.

6. Blasting of Holes in Frozen Ground. Holes are blasted in the frozen ground with the following means:

Shaped charges.
Pressure or tamped charges.

A tamped/pressure charge can be used when a layer of frozen ground is shallow. (See fig. D-9) The use of a 15lb shaped charge is also effective and will make a suitable hole for placing a charge.

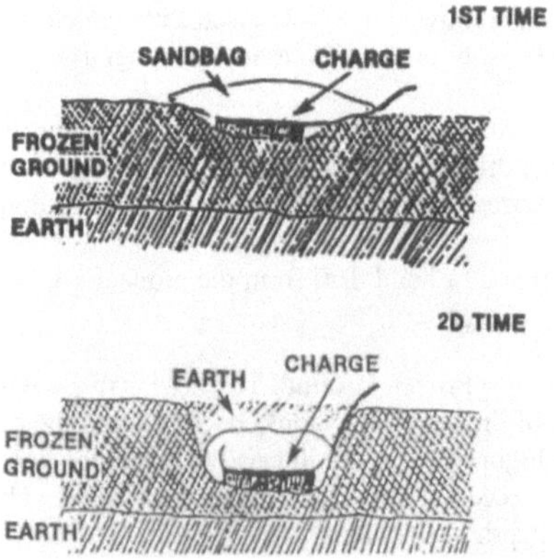

Figure B-9: Use of Pressure Charge

7. The blasting of frozen ground will demand a number of charges. With an interval between them equal to the depth of the frozen ground. If the frozen ground is 50 cm deep, fourteen holes are required, in two rows, with 50 cm intervals between the holes and 50 cm between the rows. If the frozen layer is thick, it might be necessary to repeat the blasting. The shaped charge will often pierce the layer of frozen ground. To achieve the full effect of the blast, the hole must be prepared as in figure B-10 before the second charge.

8. If a unit trench is to be made by blasting craters, one to three holes are needed. The holes are blasted through the frozen ground; then the hollow blasted beneath the frozen ground is filled in to make a suitable chamber for the main charge (refer to figure B-11). If necessary, the hole can be enlarged by the use of dynamite.

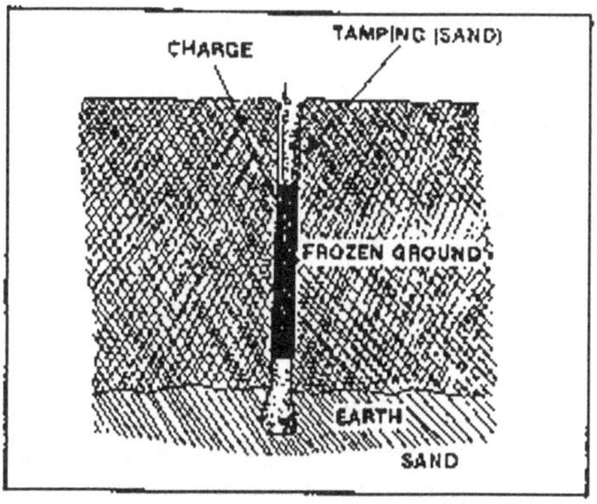

Figure B-10. Preparation of Hole Made by Shaped Charge

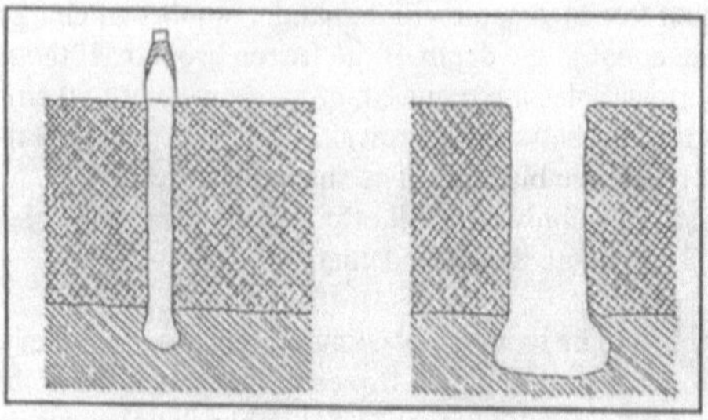

Figure B-11. Shaped Charge and Main Charge Craters

E. Ice Demolition. The purpose of blasting away ice is most often to open up a clear water gap to block the enemy. Detailed reconnaissance is necessary to get the gap opened where the current is strongest. An ice drill is needed, an axe, a tape measure, a sounded line, and the blasting equipment. The thickness of the ice and the depth of the water are measured by drilling a series of sounding holes (intervals of 50-100m). When the location of the gap has been decided, the centerline should be marked with a ski trail, twigs, or in some other manner.

1. A charge detonated underneath the ice throws a circular mound of water up, breaking the ice. If slush remains in the gap, it will freeze quickly in frigid temperatures.

2. Estimation and Placing of Charge. A gap is blasted in the ice by placing a series of charges underneath it (refer to figure B-12).

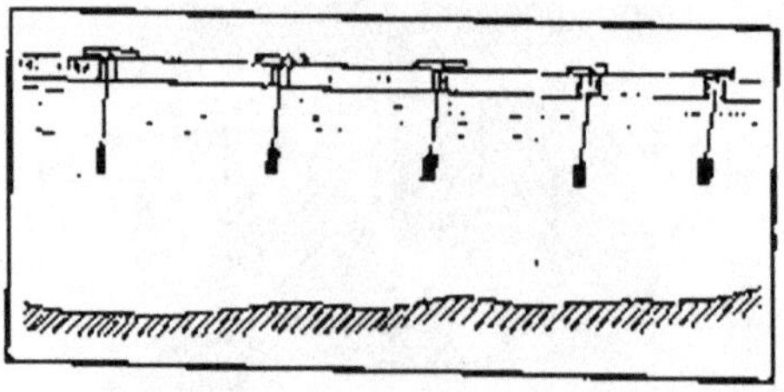

Figure B-12: Blasting a Gap in Ice

3. If the depth of the water is greater than 2.5m, the charge should be placed about 1.25 m beneath the ice. If the depth of the water is less than 2.5 m the charge should be placed at half the depth to get optimum results. The size of the charge should be as in figure B-13.

THICKNESS OF ICE	WIDTH OF THE GAP	WEIGHT OF CHARGE	INTERVALS
up to 40 cm	5 m	1 kg	4 m
up to 40 cm	6 m	2 kg	5 m
up to 40 cm	8 m	3 kg	6 m
40 – 60 cm	8 m	4 kg	8 m
60 – 100 cm	8 – 10 m	5 kg	8 m

Figure B-13: Table of Charges Depth more than 2.5 m

If the depth of water is less than 2.5 m, the hole in the ice made by any given charge becomes smaller. Therefore, the interval between the charges is reduced as shown in figure B-14

DEPTH OF WATER	INTERVAL BETWEEN CHARGES		
m	3 kg charge	4 kg charge	5 kg charg
2.0 m	5 m	7 m	9 m
1.5 m	4 m	6 m	8 m
1.0 m	4 m	5 m	6 m
0.5 m	3 m	4 m	5 m

Figure D-14. Table of Charges—Depth of Water Less Than 2.5 m.

4. TNT. C-4, and dynamite will all work as the base charge for ice breaching. Dynamite needs to be water proofed prior to use. The charges are suspended as shown in figure D-15.

5. To make a hole in the ice for the main charge, the following can be used:
   - Motor-powered ice drill.
   - Ice auger (hand for tactical purposes).
   - Explosives.
   - Axe, Pick, and crowbar.

If explosives are to be used, a hole should first be made in the ice, and the charge should be 100 grams per 10 cm of ice.

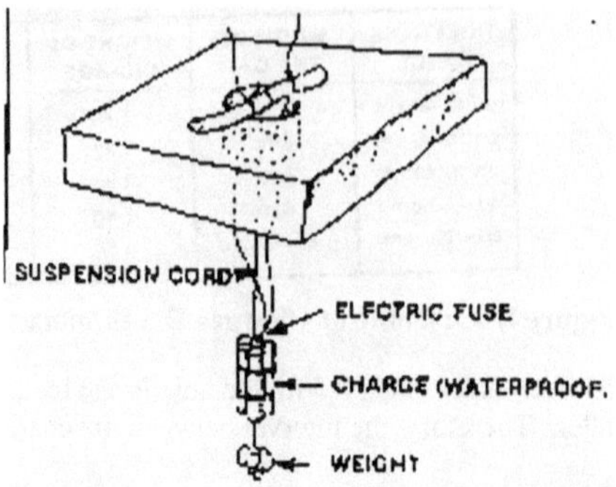

Figure B-15. Example of Suspended Charge for Demolishing Ice

6. Blasting / Breaching the Ice. The ice breaching party is organized into three teams:

    a. Reconnaissance Team. Selects appropriate site and determine the demolition requirements and report this information to the demolition team. Prepares holes as required.

    b. Demolition Team. Prepares charges in a warm, concealed position if possible. The team brings the charges on to the ice and places them properly in the holes.

    c. Initiator. Checks charge and initiate the shot. Upon detonation the team will inspect holes and check for unexploded ordnance.

7. Tactical Uses. The purpose opening gaps in the ice is to protect units against frontal and flanking attacks as well as impeding enemy travel. Breaching ice can also be used as an ambush technique against enemy patrols.

8. Water obstacles will begin to refreeze in frigid temperatures. Friendly forces must be prepared to conduct ice maintenance as needed. Covering gaps/holes with tarp or plastic sheeting can do this.

F. Artificial Obstacles. Obstacles built in a warm climate are the constructed the same with certain modifications due to the snow and frozen ground.

1. Obstacles in Frozen Ground. When timber, steel pickets, etc. cannot be driven into the ground, they can be lashed together with vertical and horizontal wraps as shown in figure D-16. When placing the obstacle on ice, construct footings for the legs, fill with snow and water to cement them in.

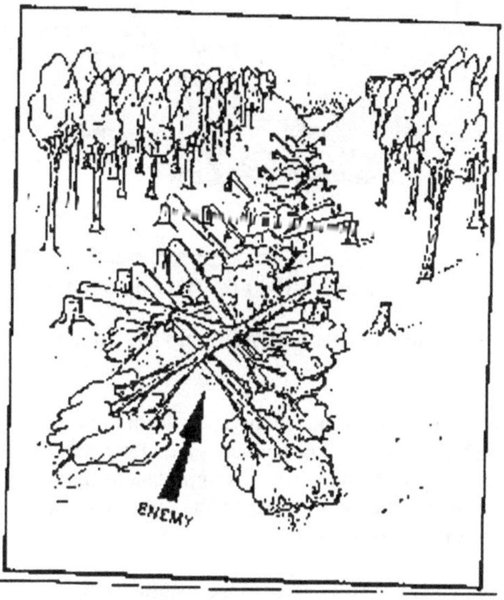

Figure B-16. Abatis.

2. Obstacles on Snow. Construct a horizontal beam and attach it to the butt ends of the obstacle or attach barbed wire or concertina to act as a floatation device. For added stability, anchor the butt ends.

3. Concertina laid on the snow (single/double/triple strand) is an easy and simple obstacle to employ against enemy skiers (refer to figure B-17).

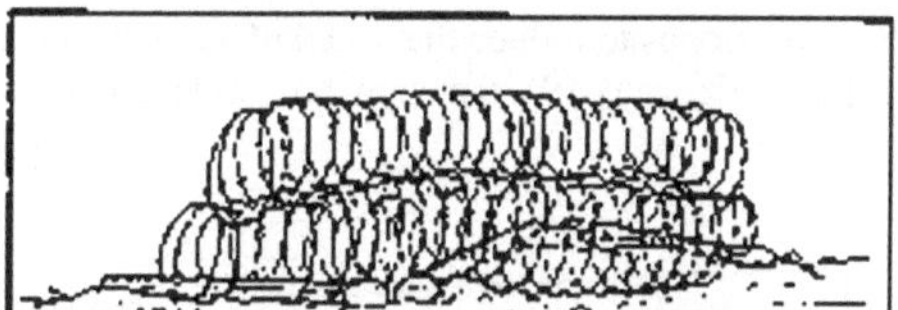

Figure B-17: Concertina on Snow

4. Using trip wires and AP mines on known enemy ski trails is an easy and effective obstacle to employ (refer to figure B-18).

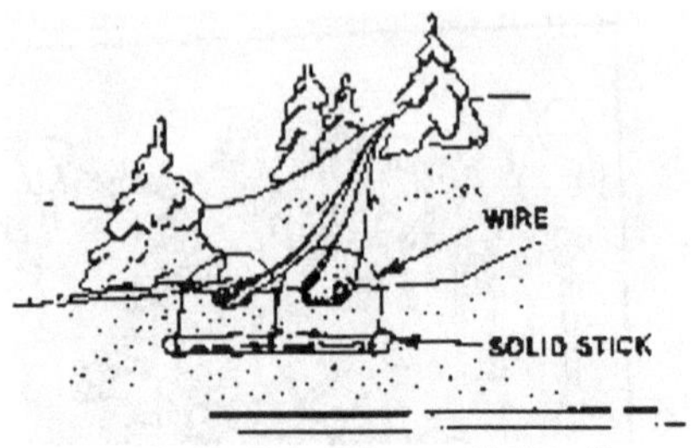

Figure B-18: Trip Wire on Ski Trail

5. Reinforce all obstacles with AP mines and trip flares to add to their effectiveness.

6. Anti-Armor, Anti-vehicle Obstacles. With the winter come added problems for vehicle travel. Snow and ice is a major problem in itself. When conducting reconnaissance, it is important to find lines of natural obstacles that require little improvement. The effort must be concentrated at the parts where the line is weakest.

7. Rivers, lakes, snow covered slopes, snow and ice, forest, and avalanche zones are just some of the natural obstacles that a winter environment brings that can impede vehicle/armor travel.

   When selecting a position, the location of natural obstacles must be carefully considered. Frequently there is more than one natural obstacle in the same place. The total effect must be carefully weighed.

8. Natural obstacles are improved when necessary:

Explosives can be used to break open waterways, create craters, lanes, and tank ditches. Obstacles can be built on slopes to reduce the speed of advance; examples of these are the timber barricade (refer to Figure B-19), the snow barricade, and felled trees such as an abatis.

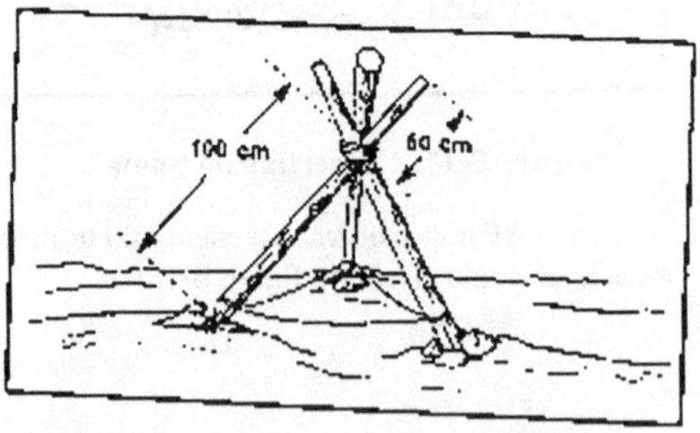

Figure B-19 Timber Obstacle

G. Mines are presently the most important passive means available for combat operations. The preparation, placing, and concealing of mines must receive particular attention in wintertime. The fuses and mechanical parts of the mine must be free of storage grease, for the grease can cause problems in very cold weather. The placing of mines in plastic bag waterproofs the mine to a degree but condensation in the bag during times of fluctuating temps can cause problems. A dark mine absorbs heat in the day and freezes at night. It is difficult to conceal mines for prolonged periods of time. The painting of the mines white could help.

1. Anti-personnel Mines. It is easy to conceal anti-personnel mines in the snow. The use of trip wires and anti-handling devices can be used but with extreme caution due to the ever-changing climate and problems that may arise with it. M605 fuses for the M16A2 should be cleaned with diesel to lubricate and preserve the fuse. Replace both safeties with safety pins to simplify the employing and removal process.

   a. All natural obstacles should be reinforced with AP mines. It is important to note that snow will soak up a great deal of the blast. All mines should be placed on a platform when employing to increase their effectiveness.

   b. A hole can be dug to the side of a ski trail. Place the mine on a platform. Slide a board under the ski track horizontally; place a stick or small piece of wood vertically on the other side of the ski track to support the horizontal board. Ensure that the prongs are in positive contact with the horizontal board, then arm and gently bury the mine (refer to figure B-20.).

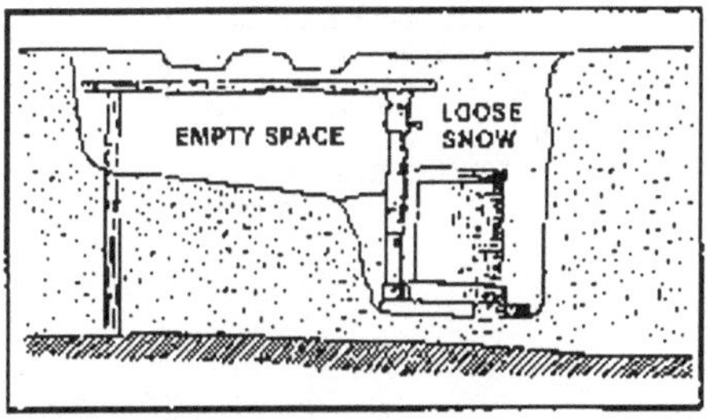

Figure B-20: Mine under Ski Trail

2. It is difficult to camouflage an anti-tank minefield in a snow-covered environment. As a rule the best place to employ an anti-tank minefield is in a well-tracked area.

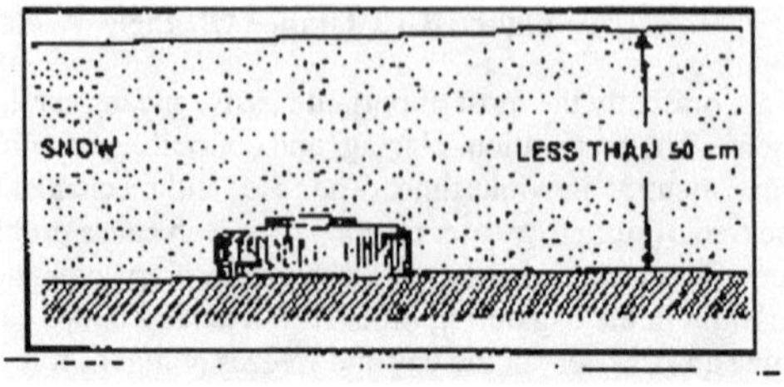

Figure B-21: Positioning the Anti-tank mine M-15

If the snow depth is 50 cm or less, placing the M-15 directly on the ground will not reduce its effectiveness. If the snow is deeper, packing the snow and using a wooden platform to rest the mine on will increase the effectiveness. A common practice is to use two mines stacking one on top of the other. The bottom is turned upside down and armed while the top is placed right side up and armed. This does two things, one it gives the top mine a platform, two the bottom mine is protected from the elements. Water seeping into the mechanical features of the bottom mine is reduced significantly. Refer to Figures B-22 and B-23.

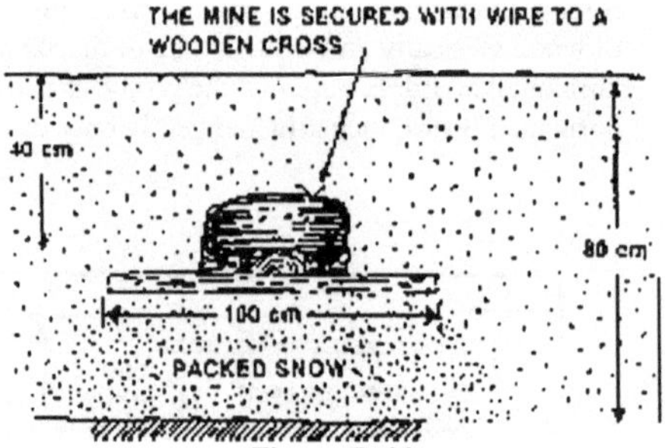

Figure B-22

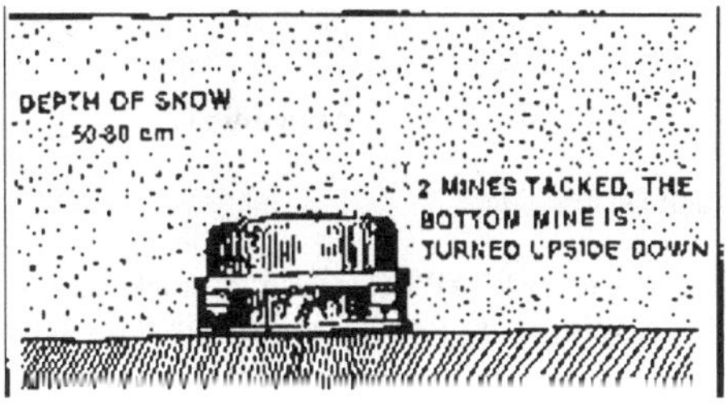

Figure B-23

Hasty surface laid minefields are still effective in this environment. They are quick, not gear or personnel intensive, and is easily employed and removed. When covered by fire they reduce the enemies' mobility heavily.

Appendix C

A. Research and Development.

1. Mattracks. Currently, MCMWTC is testing a track system made by Mattrack, that will give the HMMWV better off-road capability in snow-covered terrain. This system is designed to be installed in place of the HMMWV's wheels and consists of four 1' wide triangular rubber tracks that will improve the HMMWV's ground pressure to give the vehicle better mobility.

2. Snowmobiles. MCMWTC is developing a POI for snowmobiles to increase a unit's mobility over snow-covered terrain. Snowmobiles are being tested as administrative vehicles to run information between units, as platforms to move crew-served weapons and personnel, and increase reconnaissance units' mobility.

Made in United States
Troutdale, OR
11/03/2026

56844648R00106